GPU-BASED TECHNIQUES FOR GLOBAL ILLUMINATION EFFECTS

Synthesis Lectures on Computer Graphics and Animation

Editor
Brian A. Barsky, *University of California, Berkeley*

GPU-Based Techniques For Global Illumination Effects
László Szirmay-Kalos, László Szécsi and Mateu Sbert
2008

High Dynamic Range Imaging Reconstruction
Asla Sa, Paulo Carvalho, and Luiz Velho IMPA, Brazil
2007

High Fidelity Haptic Rendering
Miguel A. Otaduy, Ming C. Lin
2006

A Blossoming Development of Splines
Stephen Mann
2006

GPU-based Techniques for Global Illumination Effects

László Szirmay-Kalos,

László Szécsi

Mateu Sbert

www.morganclaypool.com

ISBN: 1598295594   paperback
ISBN: 9781598295597   paperback

ISBN: 1598295608   ebook
ISBN: 9781598295603   ebook

DOI 10.2200/S00107ED1V01Y200801CGR004

A Publication in the Morgan & Claypool Publishers series

SYNTHESIS LECTURES ON COMPUTER GRAPHICS AND ANIMATION #4

Lecture #4
Series Editor: Brian A. Barsky, University of California, Berkeley

Library of Congress Cataloging-in-Publication Data

Series ISSN: 1933-8996 print
Series ISSN: 1933-9003 electronic

GPU-BASED TECHNIQUES FOR GLOBAL ILLUMINATION EFFECTS

László Szirmay-Kalos

Department of Control Engineering and Information Technology,
Budapest University of Technology and Economics,
Hungary

László Szécsi

Department of Control Engineering and Information Technology,
Budapest University of Technology and Economics,
Hungary

Mateu Sbert

Institute of Informatics and Applications,
University of Girona,
Spain

SYNTHESIS LECTURES ON COMPUTER GRAPHICS AND ANIMATION #4

ABSTRACT

This book presents techniques to render photo-realistic images by programming the Graphics Processing Unit (GPU). We discuss effects such as mirror reflections, refractions, caustics, diffuse or glossy indirect illumination, radiosity, single or multiple scattering in participating media, tone reproduction, glow, and depth of field. The book targets game developers, graphics programmers, and also students with some basic understanding of computer graphics algorithms, rendering APIs like Direct3D or OpenGL, and shader programming. In order to make the book self-contained, the most important concepts of local illumination and global illumination rendering, graphics hardware, and Direct3D/HLSL programming are reviewed in the first chapters. After these introductory chapters we warm up with simple methods including shadow and environment mapping, then we move on toward advanced concepts aiming at global illumination rendering. Since it would have been impossible to give a rigorous review of all approaches proposed in this field, we go into the details of just a few methods solving each particular global illumination effect. However, a short discussion of the state of the art and links to the bibliography are also provided to refer the interested reader to techniques that are not detailed in this book. The implementation of the selected methods is also presented in HLSL, and we discuss their observed performance, merits, and disadvantages. In the last chapter, we also review how these techniques can be integrated in an advanced game engine and present case studies of their exploitation in games. Having gone through this book, the reader will have an overview of the state of the art, will be able to apply and improve these techniques, and most importantly, will be capable of developing brand new GPU algorithms.

KEYWORDS

Caustics, Depth of field, Distance map, Environment mapping, Filtering, Game engine, Games, Global illumination, GPU, Image-based lighting, Light path map, Light transport, Loose kd-tree, Obscurances, Participating media, Pre-computed radiance transfer, Radiosity, Ray engine, Ray-tracing, Real-time rendering, Reflection, Refraction, Scene graph, Shadow, Tone mapping, Virtual reality systems

Contents

List of Figures xi

Preface xxi

1. Global Illumination Rendering 1
    1.1   Materials 6
    1.2   Rendering Equation 8
    1.3   Local Illumination 10
    1.4   Global Illumination 13
    1.5   Random Walk Solution of the Rendering Equation 14
          1.5.1   Monte Carlo Integration 17
          1.5.2   Importance Sampling 19
    1.6   Iteration Solution of the Rendering Equation 20
    1.7   Application of Former GI Research Results in GPUGI 22

2. Local Illumination Rendering Pipeline of GPUS 23
    2.1   Evolution of the Fixed-function Rendering Pipeline 26
          2.1.1   Raster Operations 27
          2.1.2   Rasterization With Linear Interpolation 27
          2.1.3   Texturing 29
          2.1.4   Transformation 31
          2.1.5   Per-vertex Lighting 34
    2.2   Programmable GPUs 34
          2.2.1   Render-to-texture and Multiple Render Targets 36
          2.2.2   The Geometry Shader 36
    2.3   Architecture of Programmable GPUs 38
          2.3.1   Resources 38
          2.3.2   Pipeline Stages and Their Control by Render States 38

3. Programming and Controlling GPUS 42
    3.1   Introduction to HLSL Programming 44
          3.1.1   Vertex Shader Programming 46
          3.1.2   Geometry Shader Programming 49
          3.1.3   Fragment Shader Programming 49

3.2   Controlling the GPU from the CPU 50
    3.2.1   Controlling the Direct3D 9 Pipeline 51
    3.2.2   Controlling the Pipeline Using Effect Files 56
        3.2.2.1   Controlling the Pipeline From Direct3D 10 59
3.3   Shaders Beyond the Standard Pipeline Operation 60
    3.3.1   Are GPUs Good for Global Illumination? 60
    3.3.2   Basic Techniques for GPU Global Illumination 62

4.   Simple Improvements of the Local Illumination Model 66
4.1   Shadow Mapping 68
    4.1.1   Shadow Map Generation 70
    4.1.2   Rendering With the Shadow Map 71
    4.1.3   Shadow Map Anti-aliasing 74
4.2   Image-based Lighting 77
    4.2.1   Mirrored Reflections and Refractions 80
    4.2.2   Diffuse and Glossy Reflections Without Self-shadowing 82
    4.2.3   Diffuse and Glossy Reflections With Shadowing 85

5.   Ray Casting on the GPU 90
5.1   Ray–Triangle Intersection in a Shader 91
5.2   The Ray Engine 95
5.3   Acceleration Hierarchy Built on Rays 97
    5.3.1   Implementation of Recursive Ray-tracing Using the Ray Engine 98
    5.3.2   Ray Casting 100
5.4   Full Ray-Tracers on the GPU Using Space Partitioning 102
    5.4.1   Uniform Grid 102
    5.4.2   Octree 103
    5.4.3   Hierarchical Bounding Boxes 103
    5.4.4   Kd-Tree 104
5.5   Loose kd-Tree Traversal in a Single Shader 107
    5.5.1   GPU Representation of the Loose kd-tree 109
    5.5.2   GPU Traversal of the Loose kd-tree 111
    5.5.3   Performance 115

6.   Specular Effects with Rasterization 116
6.1   Ray-Tracing of Distance Maps 120
    6.1.1   Parallax Correction 121
    6.1.2   Linear Search 122
    6.1.3   Refinement by Secant Search 124

6.2   Single Localized Reflections and Refractions..............................126

6.3   Inter-Object Reflections and Refractions 130

6.4   Specular Reflections with Searching on the Reflector 130

6.5   Specular Reflections with Geometry or Image Transformation.............131

6.6   Self Reflections and Refractions 132

    6.6.1   Simplified Methods for Multiple Reflections and Refractions 136

6.7   Caustics...136

    6.7.1   Light Pass 137

    6.7.2   Photon Hit Filtering and Light Projection 139

6.8   Combining Different Specular Effects 144

7.   **Diffuse and Glossy Indirect Illumination** 146

7.1   Radiosity on the GPU...148

    7.1.1   Random Texel Selection 151

    7.1.2   Update of the Radiance Texture...............................153

7.2   Pre-Computed Radiosity 157

7.3   Diffuse and Glossy Final Gathering with Distance Maps.................159

8.   **Pre-computation Aided Global Illumination**............................169

8.1   Sampling..171

8.2   Finite-element Method..171

    8.2.1   Compression of Transfer Coefficients............................173

8.3   Pre-computed Radiance Transfer....................................175

    8.3.1   Direct3D Support of the Diffuse Pre-computed Radiance Transfer..179

8.4   Light Path Map 182

    8.4.1   Implementation 184

9.   **Participating Media Rendering** 195

9.1   Phase Functions 197

9.2   Particle System Model...197

9.3   Billboard Rendering...198

    9.3.1   Spherical Billboards...199

        9.3.1.1   Gouraud Shading of Particles 201

        9.3.1.2   GPU Implementation of Spherical Billboards.............202

9.4   Illuminating Participating Media 204

9.5   Rendering Explosions and Fire 205

    9.5.1   Dust and Smoke..205

    9.5.2   Fire 206

    9.5.3   Layer Composition 209

9.6   Participating Media Illumination Networks...............................210

    9.6.1   Iteration Solution of the Volumetric Rendering Equation..........211

    9.6.2   Building the Illumination Network................................212

    9.6.3   Iterating the Illumination Network..............................213

10.   **Fake Global Illumination**...218

    10.1   The Scalar Obscurances Method 220

    10.2   The Spectral Obscurances Method..................................221

    10.3   Construction of the Obscurances Map 223

    10.4   Depth Peeling 225

11.   **Postprocessing Effects**..229

    11.1   Image Filtering 229

        11.1.1 Separation of Dimensions 231

        11.1.2 Exploitation of the Bi-linear Filtering Hardware...............232

        11.1.3 Importance Sampling 234

    11.2   Glow 236

    11.3   Tone Mapping..236

        11.3.1 Local Tone Mapping 238

        11.3.2 Glow Integration Into Tone Mapping 239

        11.3.3 Temporal Luminance Adaptation 239

        11.3.4 Scotopic Vision...240

        11.3.5 Implementation 241

    11.4   Depth of Field...242

        11.4.1 Camera Models and Depth of Field 243

        11.4.2 Depth of Field With the Simulation of Circle of Confusion........244

12.   **Integrating GI Effects in Games and Virtual Reality Systems**...................248

    12.1   Game Engines and Scene Graph Managers............................249

    12.2   Combining Different Rendering Algorithms...........................253

    12.3   Case Studies..256

        12.3.1 Moria 257

        12.3.2 RT Car...259

        12.3.3 Space Station...263

Bibliography...265

List of Figures

1.1   Solid angle in which a pixel is visible by the eye.   2

1.2   Equivalence of the radiations of the display and a virtual scene.   2

1.3   Tasks of rendering   4

1.4   Color matching functions $r(\lambda)$, $g(\lambda)$, $b(\lambda)$ that express equivalent portions of red, green, and blue light mimicking monochromatic light of wavelength λ and of unit intensity. Note that these matching functions may be negative, which is the most obvious for the red matching function. Monochromatic light of wavelengths where the red color matching function is negative cannot be reproduced by adding red, green, and blue. Reproduction is only possible if the monochromatic light is mixed with some red light, i.e. some red is added to the other side, before matching.   5

1.5   Notations of the computation of radiance $L(\mathbf{x}, \omega)$ of point $\mathbf{x}$ at direction ω.   8

1.6   Phong–Blinn reflection model.   11

1.7   Comparison of local illumination rendering (left), local illumination with shadows (middle), and global illumination rendering (right). Note the mirroring torus, the refractive egg, and color bleeding on the box, which cannot be reproduced by local illumination approaches.   13

1.8   Error of the numerical quadrature in one dimension, taking $M = 5$ samples.   15

1.9   The distribution of the average with respect to the number of samples. The average has Gaussian distribution and its variance is inversely proportional to the square root of the number of samples.   18

1.10   Importance sampling.   19

1.11   Noise of the Monte Carlo methods.   20

1.12   Basis function systems and finite-element approximations.   21

2.1   Steps of incremental rendering.   24

2.2   Fixed-function pipeline.   26

2.3   Blending unit that computes the new pixel color of the frame buffer as a function of its old color (destination) and the new fragment color (source).   27

2.4   Incremental linear interpolation of property I in screen space.   28

2.5   The hardware interpolating a single property I consists of an adder ($\sum$) that increments the content of a register by a in each clock cycle. The register should store non-integers. Pixel address X is incremented by a counter to visit the pixels of a scan line. 29

2.6   Texture filtering in Direct 3D and OpenGL. 30

2.7   Shader Model 3 architecture with vertex and fragment shaders, render-to-texture, and vertex textures. 35

2.8   A chicken with per-vertex (left) and per-pixel lighting (right). Note that per-vertex lighting and linear color interpolation smears specular highlights incorrectly. This problem is solved by per-pixel lighting. 35

2.9   A typical geometry shader application: on the fly Catmull–Clark subdivision of Bay Raitt's lizard creature [3]. Normal mapping is added by the fragment shader. 37

2.10 Another typical geometry shader application: displacement mapping with correct silhouettes obtained by local ray tracing at 200 FPS on an NV8800 GPU. 37

2.11 The pipeline model of Shader Model 4.0 GPUs. 39

3.1   Structure of real-time global illumination shaders. 62

3.2   Dataflow structure of environment mapping. 62

3.3   Environment mapping using a metal shader. 63

3.4   Dataflow structure of texture atlas generation and usage. 63

3.5   A staircase model rendered to UV atlas space (left) and with the conventional 3D transformation pipeline (right). 64

3.6   Dataflow structure of deferred shading with a high number of light sources. 65

4.1   Shadow map algorithm. First the scene is rendered from the point of view of the light source and depth values are stored in the pixels of the generated image, called shadow map or depth map. Then the image is taken from the real camera, while points are also transformed to the space of the shadow map. The depth of the point from the light source is compared to the depth value stored in the shadow map. If the stored value is smaller, then the given point is not visible from the light source, and therefore it is in shadow. 68

4.2   Hardware shadow mapping with 512×512 shadow map resolution. 74

4.3   Percentage closer filtering reads the shadow map at the four corners of the lexel containing the test point, and the results of the four comparisons (either 1 or 0) are bi-linearly filtered. 75

4.4   Comparison of point sampling (left) and bi-linear filtering, i.e. 2×2 percentage closer filtering (middle), and the focusing (right) of shadow maps. 76

4.5   Comparison of classic shadow map (left) and the variance shadow map algorithm (right). We used a 5 × 5 texel filter to blur the depth and the square depth values.   77

4.6   Comparison of classic shadow map (left) and *light space perspective shadow map* [58] combined with the variance shadow map algorithm. Note that in large scenes classic shadow map makes unacceptable aliasing errors (the lexels are clearly visible) and light leak errors (the shadow leaves its caster, the column).   78

4.7   Steps of environment mapping.   79

4.8   The concept of environment mapping and the environment map of Florence [32] stored in a cube map.   80

4.9   Environment mapped reflection (left), refraction (middle), and combined reflection and refraction (right).   82

4.10   Original high dynamic range environment map (left) and the diffuse irradiance map (right).   83

4.11   Diffuse objects illuminated by an environment map.   84

4.12   Glossy objects illuminated by an environment map.   85

4.13   Radiance values of texels within the Voronoi areas are summed to compute their total power and the final Delaunay grid on high dynamic range image. The centers of Voronoi cells are the sample points.   87

4.14   Spheres in Florence rendered by taking a single sample in each Voronoi cell, assuming directional light sources (left) and using a better approximation [10] for the average reflectivity over Voronoi cells (right). The shininess of the spheres is 100.   88

4.15   Two images of a difficult scene, where only a small fraction of the environment is visible from surface points inside the room. 1000 directional samples are necessary, animation with 5 FPS is possible.   89

5.1   Nomenclature of the ray–triangle intersection calculation. Barycentric weights (b_x, b_y, b_z) identify the point at the center of mass.   92

5.2   Rendering pass implementing the ray engine.   95

5.3   Point primitives are rendered instead of full screen quads, to decompose the array of rays into tiles.   98

5.4   The rendering pass implementing the hierarchical ray engine. For every tile, the vertex buffer containing triangle and enclosing sphere data is rendered. The vertex shader discards the point primitive if the encoded triangle's circumsphere does not intersect the cone of the tile.   99

5.5   Block diagram of the recursive ray-tracing algorithm. Only the initial construction of the vertex buffer is performed on the CPU.   100

5.6   A uniform grid that partitions the bounding box of the scene into cells of uniform size. In each cell objects whose bounding box overlaps with the cell are registered.   102

5.7   Reflective objects rendered by Purcell's method at 10 FPS on 512×512 resolution using an NV6800 GPU.   103

5.8   A quadtree partitioning the plane, whose three-dimensional version is the octree. The tree is constructed by halving the cells along all coordinate axes until a cell contains "just a few" objects, or the cell size gets smaller than a threshold. Objects are registered in the leaves of the tree.   104

5.9   A kd-tree. Cells containing "many" objects are recursively subdivided into two cells with a plane that is perpendicular to one of the coordinate axes. Note that a triangle may belong to more than one cell.   105

5.10   A kd-tree cutting away empty spaces. Note that empty leaves are possible.   106

5.11   Two cell decompositions in a loose kd-tree. There is either an empty region or a shared region. Dotted triangles belong to the left cell. Contrary to classic kd-trees, triangles never extend outside their cell.   108

5.12   Construction of a loose kd-tree.   108

5.13   An example loose kd-tree and its GPU representation.   110

5.14   Traversal of a loose kd-tree branch node. We must compute the distances where the ray pierces the cutting planes to find the child ray segments. If the ray direction along the cut axis is negative, the near and far cells are reversed.   111

5.15   A scene rendered with textures and reflections (15 000 triangles, 10 FPS at 512×512 resolution).   115

6.1   Specular effects require searching for complete light paths in a single step.   117

6.2   A reflective sphere in a color box rendered by environment mapping (left) and by ray-tracing (right) for comparison. The reference point is in the middle of the color box. Note that the reflection on the sphere is very different from the correct reflection obtained by ray-tracing if the sphere surface is far from the reference point and close to the environment surface (i.e. the box).   118

6.3   A distance map with reference point $\mathbf{o}$.   119

6.4   Tracing a ray from $\mathbf{x}$ at direction $\mathbf{R}$.   120

6.5   Approximation of the hit point by $\mathbf{p}$ assuming that the surface is perpendicular to the ray direction $\mathbf{R}$.   121

6.6   Comparison of classical environment mapping and parallax correction with ray-traced reflections placing the reference point at the center of the room and moving a reflective sphere close to the environment surface.   122

6.7   Cube map texels visited when marching along ray $\mathbf{x} + R\mathbf{d}$.   123

6.8   Refinement by a secant step.   125

6.9   Left: a glass skull ($n = 1.3$, $k = 0$) of 61 000 triangles rendered at 130 FPS. Right: an aluminum teapot of 2300 triangles rendered at 440 FPS.   128

6.10   A knight in reflective armor illuminated by dynamic lights.   128

6.11   Aliasing artifacts when the numbers of linear/secant steps are maximized to 15/1, 15/10, 80/1, and 80/10, respectively.   129

6.12   Inter-object reflections in RT car game. Note the reflection of the reflective car on the beer bottles.   131

6.13   Computation of multiple refractions on a single object storing the object's normals in one distance map and the color of the environment in another distance map.   133

6.14   Single and multiple refractions on a sphere having refraction index $\nu = 1.1$.   134

6.15   Multiple refractions and reflections when the maximum ray depth is four.   134

6.16   Real caustics.   136

6.17   The light pass renders into the photon hit location image where each pixel stores the location of the photon hit.   137

6.18   The three alternatives of storing photon hit locations. Screen space methods store pixel coordinates, texture space methods texture coordinates. Cube map space methods store the direction of the photon hit with respect to the center of a cube map associated with the caustic generator (the bunny in the figure).   138

6.19   Photon hit filtering pass assuming that the photon hit location image stores texture coordinates.   140

6.20   A photon hit location image, a room rendered without blending, and the same scene with blending enabled.   142

6.21   Light projections pass assuming the the photon hit location image stores cube map texel directions.   142

6.22   Caustics seen through the refractor object.   143

6.23   Real-time caustics caused by glass objects (index of refraction $\nu = 1.3$).   143

6.24   Reflective and refractive spheres in a game environment.   144

6.25   Reflective and refractive spheres in PentaG (http://www.gebauz.com) and in Jungle-Rumble (http://www.gametools.org/html/jungle_rumble.html).   144

6.26   The Space Station game rendered with the discussed reflection, refraction, and caustic method in addition to computing diffuse interreflections (left), and compared to the result of the local illumination model (right).   145

7.1   At diffuse or glossy surfaces light paths split.   147

7.2   A single iteration step using random hemicube shooting from $\mathbf{y}_j$.   151

7.3   Random texel selection using mipmapping.   152

7.4   The two passes of the radiance update. The first pass generates a depth map to identify points visible from the shooter. The second pass transfers radiance to these points.   154

7.5   Images rendered with stochastic hemicube shooting. All objects are mapped to a single texture map of resolution 128×128, which corresponds to processing 16 000 patches.   157

7.6   Application of a light map.   158

7.7   Diffuse/glossy final gathering. Virtual lights correspond to cube map texels. These point lights are grouped to form large area lights by downsampling the cube map. At shaded point $\mathbf{x}$, the illumination of the area lights is computed without visibility tests.   160

7.8   Solid angle in which a surface is seen through a cube map pixel.   161

7.9   Notations of the evaluation of subtended solid angles.   162

7.10   A diffuse (upper row) and a glossy (lower row) skull rendered with the discussed method. The upper row compares the results of approximations in equations (7.6) and (7.7). The lower row shows the effect of not visiting all texels but only where the BRDF is maximal.   163

7.11   Diffuse bunny rendered with the classical environment mapping (left column) and with distance cube maps using different map resolutions.   165

7.12   Glossy Buddha (the shininess is 5) rendered with the classical environment mapping (left column) and with distance cube maps using different map resolutions.   166

7.13   Glossy dragon (the shininess is 5) rendered with the classical environment mapping (left column) and with distance cube maps using different map resolutions.   166

7.14   Glossy objects and a knight rendered with the distance map-based final gathering algorithms.   167

7.15   Scientist with indirect illumination obtained by the distance map-based method (left) and by the classic local illumination method (right) for comparison.   167

8.1   Light paths sharing the same light and viewing rays.   170

8.2   The basic idea of *principal component analysis*. Points in a high-dimensional (two-dimensional in the figure) space are projected onto a lower, D-dimensional subspace ($D = 1$ in the figure, thus the subspace is a line), and are given by coordinates in this low-dimensional subspace. To define these coordinates, we need a new origin $\mathbf{M}$ and basis vectors $\mathbf{B}_1, \ldots, \mathbf{B}_D$ in the lower-dimensional subspace. The origin can be the mean of original sample points. In the example of the figure there is only one basis vector, which is the direction vector of the line.   174

8.3   The simplified block diagram of Direct3D PRT. The transfer function compression results in cluster centers $\mathbf{M}^c$ and cluster basis vector $\mathbf{B}_d^c$, while lighting processing results in vector $\mathbf{L}$. In order to evaluate equation (8.4), the CPU computes scalar products $\mathbf{L} \cdot \mathbf{M}^c$ and $\mathbf{L} \cdot \mathbf{B}_d^c$ and passes them to the shader as uniform parameters. The varying parameters of the mesh vertices are transfer vector coordinates w_i^d.   180

8.4   A diffuse object rendered with PRT. In the left image only single light bounces were approximated with 1024 rays, while the right image contains light paths of maximum length 6. The pre-processing times were 12 s and 173 s, respectively. The resolution of the environment map is 256×256. The order of SH approximation is 6.   182

8.5   A diffuse object rendered with PRT. Only single light bounces were approximated with 1024 rays. The resolution of the environment map is 256×256. The order of SH approximation is 6. Pre-processing time was 723 s.   182

8.6   Overview of the preprocessing phase of the light path map method. Entry points are depicted by •, and exit points by ×. The LPM is a collection of (entry point •, exit point ×, illumination S_k) triplets, called items.   183

8.7   Overview of the rendering phase of the light path map method. The irradiance of the entry points are computed, from which the radiance of the exit points is obtained by weighting according to the LPM.   184

8.8   Notations used in the formal discussion. The CPU is responsible for building a random path starting at the entry point. The visited points of the random path are considered as virtual point light sources. The GPU computes the illumination of virtual lights at exit points.   185

8.9   Representation of an LPM as an array indexed by entry points and exit points. A single element of this map is the LPM item, a single row is the LPM pane.   189

8.10   LPM stored as 2D textures.   189

8.11   A few tiles of an LPM texture used to render Figure 8.15. Illumination corresponding to clusters of entry points is stored in tiled atlases.   190

8.12   Entry points generated randomly.   191

8.13   Comparison of local illumination and the light path map method. The lower half of these images has been rendered with local illumination, while the upper half with the light path map.   191

8.14   The chairs scene lit by a rectangular spot light. The rest is indirect illumination obtained with the light path map method at 35 FPS on an NV6800GT and close to 300 FPS on an NV8800.   192

8.15   Escher staircases scenes lit by moving lights.   192

8.16   Indirect illumination computed by the light path map method in the Space Station game (left) and the same scene rendered with local illumination (right) for comparison. Note the beautiful *color bleeding* effects produced by the light path map.   193

9.1   Modification of the radiance of a ray in participating media.   196

9.2   The lobes of Henyey–Greenstein phase function for different g values. The light direction points upward.   197

9.3   Billboard clipping artifact. When the billboard rectangle intersects an opaque object, transparency becomes spatially discontinuous.   199

9.4   Billboard popping artifact. Where the billboard crosses the front clipping plane, the transparency is discontinuous in time (the figure shows two adjacent frames in an animation).   199

9.5   Computation of length Δs the ray segment travels inside a particle sphere in camera space.   200

9.6   The accumulated density of a ray (left) and its seen opacity (right) as the function of the distance of the ray and the center in a unit sphere with constant, unit density.   201

9.7   The accumulated density of a ray as the function of the distance of the ray and the center in a unit sphere assuming that the density function linearly decreases with the distance from the particle center.   202

9.8   Particle system rendered with planar (left) and with spherical (right) billboards.   203

9.9   Final gathering for a block.   204

9.10   Comparison of participating media rendering with spherical billboards and the discussed illumination method (left) and with the classic single scattering model (right) in the Space Station game. Note that the classic method does not attenuate light correctly and exhibits billboard clipping artifacts.   205

9.11   High albedo dust and low albedo smoke.   206

9.12   Images from real smoke and fire video clips, which are used to perturb the billboard fragment opacities and temperatures.   207

9.13   Black body radiator spectral distribution.   208

9.14   Black body radiator colors from 0 K to 10,000 K. Fire particles belong to temperature values from 2500 K to 3200 K.   208

9.15   Explosion rendering algorithm.   210

9.16   Rendered frames from an explosion animation sequence.   211

9.17   Two directions of the visibility network.   213

9.18   A single particle in the illumination and visibility networks.   213

9.19   Storing the networks in arrays.   214

9.20   Notations in the fragment shader code.   216

9.21   A cloud illuminated by two directional lights rendered with different iteration steps.   216

9.22   Globally illuminated clouds of 512 particles rendered with 128 directions.   216

10.1   The cathedral rendered with the scalar obscurances algorithm and also by the standard ambient + direct illumination model (right) for comparison.   221

10.2   Comparison of spectral obscurances (right) to scalar obscurances (left). Note the color bleeding that can only be rendered with spectral obscurances.   222

10.3   A tank and a car rendered with the spectral obscurances algorithm and also by the standard ambient + direct illumination model (right) for comparison.   223

10.4   Billboard trees rendered with obscurances. In the left image the obscurances are applied to the left tree, but not to the right one to allow comparisons.   223

10.5   Digital Legend's character rendered with ambient illumination (left), obscurances map (middle), and obscurances and direct illumination (right).   224

10.6   Two snapshots of a video animating a point light source, rendered by spectral obscurances. To increase the realism the obscurance of a point is weighted by the distance from the light sources.   224

10.7   Different sampling techniques to generate cosine distributed rays.   225

10.8   Depth peeling process.   226

10.9   Six different image layers showing depth information for each pixel for the Cornell Box scene.   227

11.1   The idea of the exploitation of the bi-linear filtering hardware.   233

11.2   The glow effect.   237

11.3   The temporal adaptation process in three frames of an animation.   240

11.4   Tone mapping results.   242

11.5   Image creation of real lens.   243

11.6   Depth of field with circle of confusion.   246

11.7   The Moria game with (left) and without (right) the depth of field effect.   247

12.1   The model of shader management in scene graph software.   251

12.2   A simplified global illumination model for games.   256

12.3   A knight and a troll are fighting in Moria illuminated by the light path map algorithm.   257

12.4   A bird's eye view of the Moria scene. Note that the hall on the right is illuminated by a very bright light source. The light enters the hall on the left after multiple reflections computed by the light path map method. The troll and the knight with his burning torch can be observed in the middle of the dark hall.   258

12.5   Tone mapping with glow (left) and depth of field (right) in Moria.   258

12.6   Shaded smoke in Moria with spherical billboards and simplified multiple scattering.   259

12.7   Fireball causing heat shimmering in Moria.   259

12.8   Snapshots from the RT car game. Note that the car body specularly reflects the environment. The wheels are not only reflectors but are also caustic generators. The caustics are clearly observable on the ground. The giant beer bottles reflect and refract the light and also generate caustics. Since the car and the bottle have their own distance maps, the inter-reflections between them are also properly computed.   260

12.9   The car not only specularly but also diffusely reflects the indirect illumination, which is particularly significant when the car is in the corridor.   260

12.10   Explosions with heat shimmering in RT car demo game.   261

12.11   Indirect diffuse illumination in the Space Station game. The self illumination of the space station is computed by the light path maps method. The indirect illumination of the space station onto the scientist character is obtained by diffuse or glossy final rendering. The eyeglasses of the character specularly reflects the environment. Shadows are computed with the variance shadow map algorithm.   261

12.12   The space station rendered with scalar obscurances only.   262

12.13   A deforming glass bubble generates reflections, refractions, and caustics (upper row). The bubble is surrounded by smoke (lower row).   262

Preface

Real-time rendering poses severe computational requirements to computers. In order to maintain a continuous looking motion for the human eye, at least 20 images need to be generated in each second. An image consists of about a million pixels, which means that for a single pixel the visible point and its color should be computed in less than 50 ns in average. Note that in such a short time current CPUs are able to execute just a few tens of instructions, which are not enough to solve complex tasks. However, the identification of the point visible in a pixel requires the processing of the whole scene, which often consists of millions of objects. On the other hand, the color computation is equivalent to the simulation of the laws of electromagnetic waves and optics, and requires the solution of integral equations having complicated boundary conditions.

The required computation power can be delivered by special purpose, highly parallel hardware, which is dedicated to the particular problem of rendering. This hardware is the *Graphics Processing Unit* (*GPU*), which contains a large number of pipeline stages, and at the most computing intensive stages, the pipeline is broken to several parallel branches. As a result of the parallel architecture, in terms of floating point processing power current GPUs are equivalent to tens of current CPUs.

Parallel computation is effective if the interdependence of the processing of different data elements is minimal. Unfortunately, lighting phenomena introduce significant coupling between different elements. Surfaces may occlude each other, and the radiation of a surface point may influence the radiation of other surface points due to the light transfer and reflections. Thus, to cope with the performance requirements, the underlying physical model is often simplified. Interreflections are ignored, which limits the coupling to the solution of the visibility/occlusion problem. This simplification is called the *local illumination* model since it assumes that a point can be shaded using only its own properties, and independently of other surfaces in the scene. The local illumination model was the base of real-time rendering for many years and GPUs were built to support this scheme. However, the local illumination approach is very far from the physically accurate model of light transfer, and the simplifications of the underlying model resulted in plastic, artificial images. In reality, the color of a point indeed depends on the colors of other points, thus points cannot be shaded independently. Computation schemes taking care of these interdependences are called *global illumination* models (*GI*). Due to the gap between the computational requirements and the available hardware, computer graphics has been divided into two branches. *Real-time rendering* used local illumination approaches and were executed on the graphics hardware. Global illumination or photo-realistic images, on the other hand, were rendered on the CPU off-line, often requiring hours of computation.

A few years ago, however, the graphics hardware underwent a revolutionary change. While it kept increasing its processing power, certain stages became programmable. At the beginning of this era assembly languages were used, but nowadays there exist high level languages as well to program the GPU. Among these, the *High Level Shader Language* (*HLSL*) of the *Direct3D* environment and the *C for Graphics* (*Cg*) language for both *OpenGL* and *Direct3D* became the most popular. With these languages we are now able to modify the internal local illumination algorithm of GPUs, and force them, among others, to deliver global illumination effects. However, care should be taken since the architecture of GPUs still reflects the needs of local illumination algorithms, and we may easily lose performance if this fact is ignored. This is currently the main challenge of GPU programming.

This book presents techniques to solve various subproblems of global illumination rendering on the GPU. We assume *Direct3D 9* (also called *Shader Model 3*) or in special cases *Direct3D 10* (also called *Shader Model 4*) environments and the GPU programs are presented in HLSL. Note that due to the similarities between HLSL and Cg, most of the solutions could be easily ported to the OpenGL/Cg environment. Some important differences are highlighted at the respective sections. First, we review the global illumination rendering problem and the operation of the rendering pipeline of GPUs, then we discuss the following approaches in detail.

Shadows and image-based lighting. First, to warm up, we examine two relatively simple extensions to the local illumination rendering, *shadow mapping* and *image-based lighting* (also called *environment mapping*). Although these are not said to be global illumination methods, they definitely represent the first steps from the pure local illumination rendering toward more sophisticated global illumination approaches. These techniques already provide some insight on how the basic functionality of the local illumination pipeline can be extended with the programmable features of the GPU.

Ray-tracing. Here we present the implementation of the classic ray-tracing algorithm on the GPU. Since GPUs were designed to execute rasterization-based rendering, this approach fundamentally re-interprets the operation of the rendering pipeline.

Specular effects with rasterization. In this chapter, we return to rasterization and consider the generation of *specular effects*, including *mirror reflections*, *refractions*, and *caustics*. Note that these methods are traditionally rendered by ray-tracing, but for the sake of efficient GPU implementation, we need to generate them with rasterization.

Diffuse or glossy indirect illumination. This chapter deals with non-specular effects, which require special data structures stored in the texture memory, from which the total diffuse or glossy irradiance for an arbitrary point may be efficiently retrieved. These representations always make compromises between accuracy, storage requirements, and final gathering computation time. We present three algorithms in detail. The first one is the implementation of the *stochastic radiosity* algorithm on the GPU, which stores the radiance in a color texture. The second uses pre-computed radiosity without losing surface details. The third considers final gathering of the diffuse and glossy indirect illumination using *distance cube maps*.

Pre-computation aided global illumination. These algorithms pre-compute the effects of light paths and store these data compactly in the texture memory for later reuse. Of course, pre-computation is possible if the scene is static. Then during the real-time part of the process, the actual lighting is combined with the prepared data and real-time global illumination results are provided. Having presented the theory of the *finite-element method* and *sampling*, we discuss *pre-computed radiance transfer (PRT)* using finite-element representation, and the *light path map* method that is based on sampling.

Participating media or *volumetric models* may scatter light not only on the surfaces but inside the object as well. Such media are described by the *volumetric rendering equation*, which can also be solved on the GPU. We consider *particle systems* and their artifact-free rendering, then a simplified multiple scattering model is presented.

Fake global illumination. There are methods that achieve high frame rates by simplifying the underlying problem. These approaches are based on the recognition that global illumination is inherently complex because the illumination of every point may influence the illumination of every other point in the scene. However, the influence diminishes with the distance, thus it is worth considering only the local neighborhood of each point during shading. Methods using this simplification include *obscurances* and *ambient occlusion*, from which the first one is presented in detail. Note that these methods are not physically plausible, but provide satisfying results in many applications.

Postprocessing effects. Global illumination computations result in *high dynamic range*, i.e. floating point images, which need to be converted to unsigned bytes for conventional displays. This conversion takes into account the adaptation of the human eye or the settings of the camera, and is called *tone mapping*. The conversion process may also simulate other camera effects such as *glow* or *depth of field*.

When the particular methods are discussed, images and rendering times are also provided. If it is not stated explicitly, the performance values (e.g. *frames per second* or *FPS* for short) have been measured on an Intel P4 3 GHz PC with 1GB RAM and NVIDIA GeForce 6800 GT graphics card in full screen mode (1280 × 1024 resolution).

In this book, we tried to use unified notations when discussing different approaches. The most general notations are also listed here.

a: the *albedo* of a material.

$D^2(\xi)$: the *variance* of random variable ξ.

DepthWorldViewProj: a uniform shader program parameter of type float4x4, which defines the transformation matrix from modeling space to clipping space used when rendering the depth map.

DepthWorldViewProjTex: a uniform shader program parameter of type float4x4, which defines the transformation matrix from modeling space to texture space of the depth map.

$E(\xi)$: the *expected value* of random variable ξ.

EyePos: a uniform shader program parameter of type float3, which defines the camera position in world space.

$f_r(\omega', \mathbf{x}, \omega)$: the BRDF function at point $\mathbf{x}$ for the illumination direction ω' and the viewing direction ω. If the surface is diffuse, the BRDF is denoted by $f_r(\mathbf{x})$.

$F(\theta')$: the Fresnel function when the illumination direction encloses angle θ' with the surface normal.

$G(\mathbf{x}, \mathbf{y})$: the *geometric factor* between two points.

$\mathbf{H}$: the unit length *halfway vector* between the illumination and viewing directions.

I: the *irradiance*.

$\mathbf{L}$: the unit length *illumination direction*.

$L(\mathbf{x}, \omega)$: the *radiance* of point $\mathbf{x}$ at direction ω.

$L^e(\mathbf{x}, \omega)$: the *emitted radiance* of point $\mathbf{x}$ at direction ω.

$L^{\mathrm{env}}(\omega)$: the radiance of the environment illumination from direction ω.

$L^r(\mathbf{x}, \omega)$: the *reflected radiance* of point $\mathbf{x}$ at direction ω.

$\mathbf{N}$: the unit length *normal vector*.

Norm, wNorm, cNorm: vertex or fragment normals in modeling, world, and camera spaces, respectively.

$\mathbf{o}$: the center or *reference point* of a cube map.

oColor, oTex: vertex shader output color and texture coordinates.

Pos, wPos, cPos, hPos: vertex or fragment positions in modeling, world, camera, and clipping spaces, respectively.

S: the set of all surface points.

$v(\mathbf{x}, \mathbf{y})$: *visibility indicator* which is 1 if points $\mathbf{x}$ and $\mathbf{y}$ are visible from each other and zero otherwise.

$\mathbf{V}$: the unit length *viewing direction*.

`World`: a uniform shader program parameter of type `float4x4`, which transforms from modeling space to world space.

`WorldIT`: a uniform shader program parameter of type `float4x4`, the inverse transpose of `World`, and is used to transform normal vectors from modeling space to world space.

`WorldView`: a uniform shader program parameter of type `float4x4`, which defines the transformation matrix for place vectors (points) from modeling space to camera space.

`WorldViewIT`: a uniform shader program parameter of type `float4x4`, which defines the inverse transpose of `WorldView`, and is used to transform normal vectors from modeling space to camera space.

`WorldViewProj`: a uniform shader program parameter of type `float4x4`, which defines the transformation matrix from modeling space to clipping space.

x: the point to be shaded, which is the *receiver* of the illumination.

y: the point that is the *source* of the illumination.

Y: the *luminance* of a spectrum.

Greek Symbols

θ: the angle between the viewing direction and the surface normal.

θ': the angle between the illumination direction and the surface normal.

ω: the *outgoing direction* or *viewing direction*.

ω': the *incoming direction* or *illumination direction*.

Ω: set of all outgoing directions.

Ω': the set of all incoming directions.

ACKNOWLEDGEMENTS

The research work that resulted in this book has been supported by the GameTools FP6 (IST-2-004363) project, Hewlett-Packard, the National Office for Research and Technology (Hungary), OTKA (T042735), and TIN2007-68066-C04-01 of the Spanish Government. In this book, we used many images created during the GameTools project by Tamás Umenhoffer, Balázs Tóth, Nicolau Sunyer, István Lazányi, Barnabás Aszódi, György Antal, Ismael Garcia, Xavier Carrillo, Kristóf Ralovich, Attila Barsi, Michael Wimmer, Markus Giegl, and others. We thank them for their contribution and also for the helpful discussions that improved the text of this book.

The authors

CHAPTER 1

Global Illumination Rendering

The ultimate objective of *image synthesis* or *rendering* is to provide the user with the illusion of watching real objects on the computer screen. The image on the computer screen consists of constant color small rectangles, called *pixels*. The color of pixels is generated from an internal model that is called the *virtual world* or the *scene*. To provide the illusion of watching the real world, the color sensation of an observer looking at the artificial image generated by the graphics display must be approximately equivalent to the color perception which would be obtained in the real world. The color perception of humans depends on the *power* spectrum of the light arriving at the eye from different viewing directions, i.e. on the number of photons of different frequencies. If we were able to get the computer display to emit photons of the same number and frequency as the virtual world, then the eye would not be able to distinguish the displayed image from the real world.

To examine this *visual equivalence* formally for a single pixel of the image, we use the notations of Figure 1.1. The area of the considered pixel is ΔA_1, the distance between its center and the center of the human eye is r_1. The points of the pixel are seen from the eye in directions belonging to the *solid angle* subtended by the pixel. If we look at a surface at grazing angles or the surface is farther away, the surface area seems to be smaller, i.e. its subtended solid angle is smaller. More formally, solid angle $\Delta\omega$ subtended by small surface ΔA_1 at distance r_1 is $\Delta A_1 \cos\theta_1 / r_1^2$, where θ_1 is the angle between the surface normal and the viewing direction.

Let the solid angle subtended by the pupil from the center of the pixel be $\Delta\omega_1$ (left of Figure 1.2). Those photons emitted by the pixel affect the eye which leave the pixel at a direction in the solid angle $\Delta\omega_1$. Let us denote the power in a given wavelength range $[\lambda, \lambda + d\lambda)$ transported from the pixel of area ΔA_1 to the eye by $\Phi_\lambda(\Delta A_1, \Delta\omega_1)$.

A pixel represents that part of the virtual world that is seen in it (right of Figure 1.2). The surface area visible in this pixel is denoted by ΔA_2, its distance from the eye by r_2, and the solid angle in which the pupil is visible from the center of this small surface area by $\Delta\omega_2$. Since the pixel and the surface subtend the same solid angle, $\Delta\omega$, we obtain the following relation between the areas, distances, and orientation angles of the pixel and the

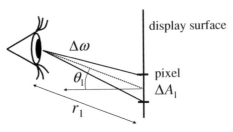

FIGURE 1.1: Solid angle in which a pixel is visible by the eye.

visible surface, respectively:

$$\Delta\omega = \frac{\Delta A_1 \cos\theta_1}{r_1^2} = \frac{\Delta A_2 \cos\theta_2}{r_2^2}. \tag{1.1}$$

The power in $[\lambda, \lambda + d\lambda)$ transported from the visible surface to the eye is $\Phi_\lambda(\Delta A_2, \Delta\omega_2)$. The visual equivalence of the pixel and the surface visible in it can be expressed by the following power identity:

$$\Phi_\lambda(\Delta A_1, \Delta\omega_1) = \Phi_\lambda(\Delta A_2, \Delta\omega_2).$$

This requirement is difficult to meet since the area visible in a pixel (ΔA_2) depends on the distance from the eye and also on the orientation of the surface. The solid angles subtended by the pupil from the pixel $(\Delta\omega_1)$ and from the visible surface $(\Delta\omega_2)$ are also difficult to obtain. Fortunately, a much simpler identity can be found for the visual equivalence if the pixel is small (which is usually the case).

Let us realize that the solid angles $\Delta\omega_1$ and $\Delta\omega_2$ in which the pupil is seen from the pixel and from the visible surface are inversely proportional to the square of the distance,

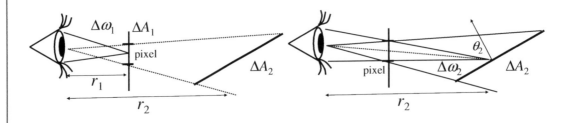

FIGURE 1.2: Equivalence of the radiations of the display and a virtual scene.

that is,

$$\frac{\Delta\omega_1}{\Delta\omega_2} = \frac{r_2^2}{r_1^2}.$$

Multiplying this with equation (1.1), the dependence on the distances r_1 and r_2 can be eliminated since the two factors compensate each other, thus we obtain the following geometric identity:

$$\Delta A_1 \cos\theta_1 \Delta\omega_1 = \Delta A_2 \cos\theta_2 \Delta\omega_2.$$

Dividing the power identity of the visual equivalence by this geometric identity, we conclude that the requirement of visual equivalence is the similarity of the power density computed with respect to the area and the solid angle:

$$\frac{\Phi_\lambda(\Delta A_1, \Delta\omega_1)}{\Delta A_1 \cos\theta_1 \Delta\omega_1} = \frac{\Phi_\lambda(\Delta A_2, \Delta\omega_2)}{\Delta A_2 \cos\theta_2 \Delta\omega_2}.$$

This power density is called the *radiance*:

$$L_\lambda(\mathbf{x}, \omega) = \frac{\Phi_\lambda(\Delta A, \Delta\omega)}{\Delta A \cos\theta \Delta\omega}. \qquad (1.2)$$

If two surfaces have the same radiance on all wavelengths, then the human eye perceives them as having the same color no matter how far the two surfaces are. Since our objective is to mimic the real world color sensation, the radiance of the virtual world should be computed and the pixels of the monitor should be controlled accordingly. The process of computing the radiance of the points visible through the pixels is called *rendering*. Since the radiance characterizes the brightness and the color of a surface point independently of the distance of the observer, the brightness and the color remain the same while we are going farther from a surface. The greater the distance, the larger the surface area that is visible in the solid angle measured by a single photopigment of our eye, and the two factors compensate each other. This is not true for very small or far objects, which are visible in a solid angle that is smaller than the solid angle measured by a single photopigment. The perceived radiance of such small objects diminish with the square of the distance. The most typical example of such small objects is the *point light source*.

Rendering takes the virtual world, including its *geometry*, *optical material properties*, and the description of *lighting*, and computes the radiance of the surface points applying the laws of physics on the wavelengths of the spectrum visible by the human eye (Figure 1.3). The exact simulation of the light perceived by the eye is impossible, since it would require an endless computational process. On the other hand, it is not even worth doing since the possible spectral power distributions which can be produced by computer screens are limited in contrast to the infinite variety of real world light distributions. Consequently, color perception

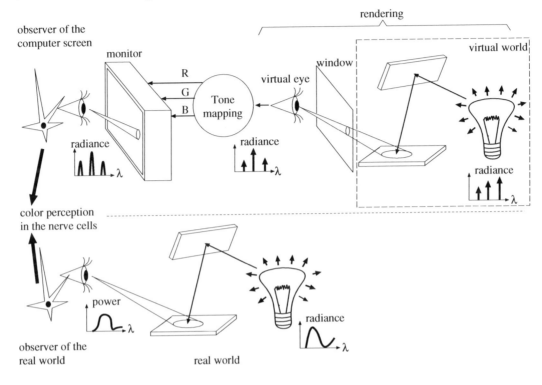

FIGURE 1.3: Tasks of rendering.

is approximated instead of having a completely accurate simulation. The accuracy of this approximation is determined by the ability of the eye to make a distinction between two spectral power distributions. The human eye actually samples and integrates the energy in three overlapping frequency ranges by three types of *photopigments*, which means that the color sensation of an arbitrary spectrum can be made similar to the sensation of a very different spectrum delivering energy just on three wavelengths that are far enough from each other. These wavelengths usually correspond to the red, green, and blue colors, also called *tristimulus coordinates* or *additive color primaries*. A possible choice for the wavelengths is

$$\lambda_{\text{red}} = 645 \text{ nm}, \qquad \lambda_{\text{green}} = 526 \text{ nm}, \qquad \lambda_{\text{blue}} = 444 \text{ nm}. \qquad (1.3)$$

The equivalent portions of red, green, and blue light mimicking monochromatic light of wavelength λ and of unit intensity are shown by the *color matching functions* ($r(\lambda)$, $g(\lambda)$, and $b(\lambda)$) of Figure 1.4. These functions are based on physiological measurements.

   If the perceived color is not monochromatic, but is described by an L_λ distribution, the tristimulus coordinates are computed using the assumption that the sensation is produced by

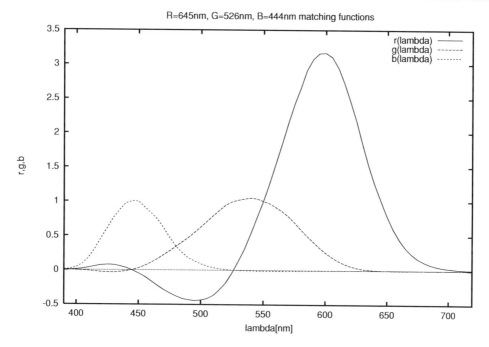

FIGURE 1.4: Color matching functions $r(\lambda)$, $g(\lambda)$, $b(\lambda)$ that express equivalent portions of red, green, and blue light mimicking monochromatic light of wavelength λ and of unit intensity. Note that these matching functions may be negative, which is the most obvious for the red matching function. Monochromatic light of wavelengths where the red color matching function is negative cannot be reproduced by adding red, green, and blue. Reproduction is only possible if the monochromatic light is mixed with some red light, i.e. some red is added to the other side, before matching.

an additive mixture of the perceptions of elemental monochromatic components:

$$R = \int_\lambda r(\lambda)L_\lambda \, d\lambda, \qquad G = \int_\lambda g(\lambda)L_\lambda \, d\lambda, \qquad B = \int_\lambda b(\lambda)L_\lambda \, d\lambda. \qquad (1.4)$$

Thus in the final step of image synthesis a *tone mapping* process is needed, which converts the computed radiance spectrum to displayable red, green, and blue intensities. The calculation of displayable red, green, and blue values involves the evaluation of integrals of equation (1.4) in the general case, which is called *spectral rendering*. The color matching process is quite time consuming, thus we usually take simplifications. The radiance computation is carried out just on the representative wavelengths of red, green, and blue, and the results directly specify the (R, G, B) pixel intensities. In this *RGB rendering* method, the tone mapping is only responsible for scaling the computed red, green, and blue to values that can be written into the color buffer and displayed by the monitor. It is only an approximation, but usually provides acceptable results. In this book, we shall generally assume that the radiance is computed just on these three wavelengths. From this point we shall not include wavelength λ in the notation of radiance L. When the radiance symbol shows up in an equation, we should think of three similar equations, one for each of the representative wavelengths of red, green, and blue.

Colors are computed with some accuracy, which might be more or less obvious to the observer. When these artifacts are needed to be reduced, we may take into account that the sensitivity of the human eye varies with the wavelength. We can assign a scalar *luminance* value to a spectrum that describes how bright the surface appears, i.e. its perceived intensity. The luminance can also be expressed by the R, G, B primaries. If the primaries correspond to the *CIE* standard, then the transform is

$$Y = 0.2126R + 0.7152G + 0.0722B.$$

1.1   MATERIALS

The radiance of a light ray is constant until it collides with some object of a given material. The radiance of the light after scattering is proportional to the *incident radiance* and also depends on the material properties.

If the surface is smooth, then the light may get reflected into the *ideal reflection direction* specified by the *reflection law*[1] of the geometric optics, and may get refracted into the refraction direction according to the *Snellius–Descartes law of refraction*[2]. Incident radiance L^{in} coming from a direction that encloses the *incident angle* θ' with the surface normal is broken down to *reflected radiance* L^r and *refracted radiance* L^t according to the *Fresnel function* $F(\theta')$, which depends both on the wavelength and on the incident angle:

$$L^r = L^{in} F(\theta'), \quad L^t = L^{in}(1 - F(\theta')).$$

The Fresnel function defines the probability that a photon is reflected when it collides with an optically smooth surface. Using the Maxwell equations, the Fresnel function can be expressed from the incident angle θ', index of refraction ν, and *extinction coefficient* κ that describes how

---

[1]The reflection law states that the illumination direction, reflection direction, and the surface normal of the reflector are in the same plane, and the incident angle between the surface normal and the illumination direction equals the outgoing angle that is between the surface normal and the reflection direction.
[2]The refraction law states that the illumination direction, refraction direction, and the surface normal of the refractor surface are in the same plane, and the sine of the incident angle θ', that is between the surface normal and the illumination direction, and the sine of refraction angle θ, that is between the surface normal and the refraction direction, satisfy $\sin\theta'/\sin\theta = \nu$, where ν is the relative index of refraction, which equals the ratio of the speeds of light in the two materials separated by the refractor surface.

quickly the light diminishes inside the material (a rough guess for the extinction coefficient is the wavelength divided by the mean traveled distance):

$$F(\theta') = \frac{1}{2}\left(\left|\frac{\cos\theta' - (\nu + \kappa j)\cos\theta}{\cos\theta' + (\nu + \kappa j)\cos\theta}\right|^2 + \left|\frac{\cos\theta - (\nu + \kappa j)\cos\theta'}{\cos\theta + (\nu + \kappa j)\cos\theta'}\right|^2\right), \qquad (1.5)$$

where $j = \sqrt{-1}$, and θ is the refraction angle. The first term corresponds to the case when the *polarization* of the light is parallel with the surface (i.e. light's electric vector oscillates in a plane that is parallel with the reflector), while the second term describes the reflection of perpendicularly polarized light. If the light is not polarized, the Fresnel factor is the average of these two cases as shown in equation (1.5). Since the exact Fresnel formula is too complex, we usually apply an approximation [85, 123]

$$F(\theta') = F_\perp + (1 - F_\perp)(1 - \cos\theta')^5,$$

where

$$F_\perp = \frac{(\nu - 1)^2 + \kappa^2}{(\nu + 1)^2 + \kappa^2} \qquad (1.6)$$

is the Fresnel function (i.e. the probability that the photon is reflected) at perpendicular illumination. Note that $F_\perp$ is constant for a given material on a given wavelength, thus this value can be computed once and passed to the shader as a global variable.

Real surfaces are not ideally smooth but can be modeled as collections of randomly oriented smooth *microfacets*. The theory discussed so far and the Fresnel function describe these rough surfaces just on the level of these microfacets. A photopigment in the human eye "sees" a collection of these microfacets, thus perceives their aggregate behavior. On the human scale, the surface may reflect the incident light not only in a single direction but in infinitely many directions. The material is described by its *Bi-directional Reflection/refraction Function* (*BRDF*), which can be derived from the Fresnel function and the probabilistic model of the microfacet structure. On a single wavelength, the BRDF is the ratio of the reflected radiance L^r and the product of the incident radiance L^{in} and cosine of incident angle θ' that is between the illumination direction and the surface normal:

$$f_r(\omega', \mathbf{x}, \omega) = \frac{L^r}{L^{in}\cos\theta'}.$$

The inclusion of the cosine of the incident angle in the definition of the BRDF can be justified if we consider that an incident light ray arriving at grazing angles illuminates a surface area that is larger than its cross section, thus the light density on the surface will be proportional to this cosine factor. Recognizing this, we often use *irradiance* I instead of the *incident radiance* L^{in}, which is defined as $I = L^{in}\cos\theta'$. On the other hand, the BRDF defined by the ratio of the

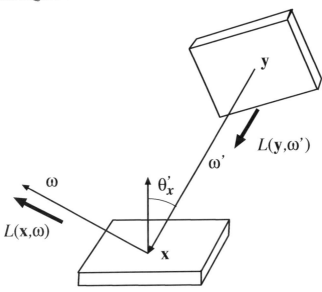

FIGURE 1.5: Notations of the computation of radiance $L(\mathbf{x}, \omega)$ of point $\mathbf{x}$ at direction ω.

reflected radiance and the irradiance is symmetric for real materials, i.e. illumination direction ω' can be exchanged with the viewing direction ω in the BRDF formula. This relation is called *Helmholtz reciprocity*.

1.2   RENDERING EQUATION

Rendering algorithms evaluate the radiance spectrum at a few representative wavelengths, most conveniently at the wavelengths of red, green, and blue since this relieves us from the color matching calculation. In scenes not incorporating *participating media*, it is enough to calculate the radiance at surface points since in vacuum the transferred radiance does not change along a light ray. The radiance reflected off a surface is affected by the emission of this point, the illumination provided by other surfaces, and the optical properties of the material at this point.

Formally this dependence is characterized by a Fredholm-type integral equation, which is called the *rendering equation* (Figure 1.5). Denoting the radiance of point $\mathbf{y}$ in direction ω' by $L(\mathbf{y}, \omega')$, the rendering equation expresses radiance $L(\mathbf{x}, \omega)$ at point $\mathbf{x}$ in direction ω as its own *emission* $L^e(\mathbf{x}, \omega)$ and the sum of contributions scattered from all incoming directions ω':

$$L(\mathbf{x}, \omega) = L^e(\mathbf{x}, \omega) + \int_{\Omega'} L(\mathbf{y}, \omega') f_r(\omega', \mathbf{x}, \omega) \cos^+ \theta'_{\mathbf{x}} \, d\omega', \qquad (1.7)$$

where $\mathbf{y}$ is the point visible from $\mathbf{x}$ at direction ω', Ω' is the directional sphere, $f_r(\omega', \mathbf{x}, \omega)$ is the *BRDF*, and $\theta'_{\mathbf{x}}$ is the incident angle between the surface normal and direction $-\omega'$ at $\mathbf{x}$. If the incident angle $\theta'_{\mathbf{x}}$ is greater than 90 degrees—i.e. the light illuminates the "back" of the surface—then the negative cosine value should be replaced by zero, which is indicated by superscript $^+$ in $\cos^+$. Illuminating point $\mathbf{y}$ is unambiguously determined by the shaded point $\mathbf{x}$ and the illumination direction ω'. This dependence is often expressed by the *ray-tracing function*

$$\mathbf{y} = h(\mathbf{x}, -\omega'),$$

which is responsible for the representation of the geometry in the rendering equation.

The rendering equation describes the light transfer at a single representative wavelength. In fact, we always solve one such equation for each representative wavelength.

The integral of the rendering equation computes a single reflection of the radiance field $L(\mathbf{y}, \omega')$ and generates its reflected radiance $L^r(\mathbf{x}, \omega)$ at every point $\mathbf{x}$ and direction ω, and can be regarded as the *transport operator* $\mathcal{T}_{f_r}$ that takes the radiance function $L(\mathbf{y}, \omega')$ and computes its reflection $L^r(\mathbf{x}, \omega)$:

$$L^r(\mathbf{x}, \omega) = \mathcal{T}_{f_r} L = \int_{\Omega'} L(\mathbf{y}, \omega') f_r(\omega', \mathbf{x}, \omega) \cos^+ \theta'_{\mathbf{x}} \, d\omega'.$$

The rendering equation integrates in the domain of input directions Ω'. If a differential surface dy is visible from illuminated point $\mathbf{x}$, and the angle between direction $\omega'_{\mathbf{y} \to \mathbf{x}}$ from $\mathbf{y}$ to $\mathbf{x}$ and the surface normal at $\mathbf{y}$ is $\theta_{\mathbf{y}}$, then this differential surface appears in solid angle

$$d\omega' = \frac{dy \cos^+ \theta_{\mathbf{y}}}{|\mathbf{x} - \mathbf{y}|^2},$$

that is, it is proportional to the size of the illuminating surface, it depends on its orientation, and inversely proportional to the square of the distance. If the differential surface dy is occluded, then the solid angle $d\omega'$ is zero. To model visibility, we can introduce a *visibility indicator* $v(\mathbf{x}, \mathbf{y})$ that is 1 if the two points see each other and zero otherwise. With these, an equivalent form of the transport operator uses an integration over the *set of surface points* S:

$$L^r(\mathbf{x}, \omega) = \int_S L(\mathbf{y}, \omega'_{\mathbf{y} \to \mathbf{x}}) f_r(\omega'_{\mathbf{y} \to \mathbf{x}}, \mathbf{x}, \omega) G(\mathbf{x}, \mathbf{y}) \, dy. \qquad (1.8)$$

In this equation

$$G(\mathbf{x}, \mathbf{y}) = v(\mathbf{x}, \mathbf{y}) \frac{\cos^+ \theta'_{\mathbf{x}} \cos^+ \theta_{\mathbf{y}}}{|\mathbf{x} - \mathbf{y}|^2} \qquad (1.9)$$

is the *geometric factor*, where $\theta'_{\mathbf{x}}$ and $\theta_{\mathbf{y}}$ are the angles between the surface normals and direction $\omega_{\mathbf{y} \to \mathbf{x}}$ that is between $\mathbf{y}$ and $\mathbf{x}$.

If we knew the radiance of all other points $\mathbf{y}$, then the solution of the rendering equation would be equivalent to the evaluation of an integral. Unfortunately, the radiance of other points is not known, but similar equations should be solved for them. In fact, the unknown radiance function shows up not only on the left side of the rendering equation, but also inside the integral of the right side. The rendering equation is thus an integral equation. Solving integral equations is difficult and requires a lot of computation.

1.3   LOCAL ILLUMINATION

Local illumination methods take a drastic approach, and approximate the radiance to be reflected by a known term, for example, by the emission of the light sources. Formally, these methods evaluate the following simplified rendering equation to obtain the radiance:

$$L(\mathbf{x}, \omega) = L^e(\mathbf{x}, \omega) + \int_{\Omega'} L^{\text{in}}(\mathbf{x}, \omega') f_r(\omega', \mathbf{x}, \omega) \cos^+ \theta'_{\mathbf{x}} \, d\omega', \qquad (1.10)$$

where $L^{\text{in}}(\mathbf{x}, \omega')$ is some approximation of the incoming radiance at point $\mathbf{x}$. *Abstract light sources*, such as *point* or *directional light sources* are preferred here, since they can provide illumination just in a single direction for each point, which simplifies the integral of equation (1.10) to a sum of terms for each abstract light source:

$$L(\mathbf{x}, \omega) = L^e(\mathbf{x}, \omega) + \sum_l L_l^{\text{in}}(\mathbf{x}) f_r(\omega'_l, \mathbf{x}, \omega) \cos^+ \theta'_l, \qquad (1.11)$$

where l runs for each of the abstract light sources, $L_l^{\text{in}}(\mathbf{x})$ is the incoming radiance at point $\mathbf{x}$ due to light source l, ω'_l is the direction of the incoming illumination, and θ'_l is the angle between the surface normal at $\mathbf{x}$ and the direction of the illumination. The direction of the illumination is independent of the shaded point for directional lights. For point light sources, the illumination direction points from the light source position to the illuminated point. If the light sources are assumed to be visible everywhere, then $L_l^{\text{in}}(\mathbf{x})$ is constant for directional light sources and is inversely proportional to the distance between the shaded point and the location of the light source for point light sources. Let us denote the unit length surface normal by vector $\mathbf{N}$ and the unit length illumination direction parallel to $-\omega'_l$ by vector $\mathbf{L}$. The cosine term can be obtained by a simple dot product of the two unit vectors:

$$\cos^+ \theta'_l = (\mathbf{N} \cdot \mathbf{L})^+.$$

Again, superscript $^+$ indicates that negative dot products are replaced by zero.

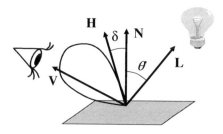

FIGURE 1.6: Phong–Blinn reflection model.

In the very popular *Phong–Blinn reflection model*, the reflection is supposed to be the sum of a diffuse reflection and of a glossy reflection. For *diffuse reflection*, the BRDF is independent of the incoming and outgoing directions, thus, it can be represented by the constant k_d on a single wavelength. For *glossy reflection*, the product of the BRDF and $\cos^+ \theta_l'$ are defined by a function, which is maximal when the viewing direction is equal to the ideal reflection direction and decreases as the viewing direction gets farther from it (Figure 1.6). In order to define the "distance" between unit length viewing direction **V** and the reflection direction of **L**, their unit *halfway vector* is found as

$$\mathbf{H} = \frac{\mathbf{V} + \mathbf{L}}{|\mathbf{V} + \mathbf{L}|},$$

and the cosine of the angle between this halfway vector and the unit normal is computed:

$$\cos^+ \delta = (\mathbf{N} \cdot \mathbf{H})^+.$$

Note that if the viewing vector **V** were just the reflection of **L** onto the unit normal **N**, then the halfway vector **H** would be identical to the normal. In this case, δ is zero and its cosine is maximal. To control the effect of this angle on the reflected intensity, that is to express how reflective and how polished the material is, the product of the *glossy reflectivity* k_s and the *n*th power of this cosine is used where *n* expresses the *shininess* of the surface. Putting these together, the Phong–Blinn formula is

$$f_r(\mathbf{L}, \mathbf{x}, \mathbf{V})(\mathbf{N} \cdot \mathbf{L})^+ = k_d(\mathbf{N} \cdot \mathbf{L})^+ + k_s \left((\mathbf{N} \cdot \mathbf{H})^+ \right)^n.$$

In this formula, reflectivities k_d and k_s are wavelength dependent.

The local illumination model examines only one-bounce light paths and ignores multiple reflections. This clearly results in some illumination deficit, so the images will be darker than expected. More importantly, those surfaces that are not directly visible from any light sources will be completely dark. The old trick to add the missing illumination in a simplified form is the introduction of the *ambient light*. Let us assume that there is some ambient lighting in the scene of intensity L^a that is constant for every point and every direction. According to the rendering equation (1.7), the reflection of this ambient illumination in point $\mathbf{x}$ and at viewing direction ω is

$$L(\mathbf{x}, \omega) = \int_{\Omega'} L^a f_r(\omega', \mathbf{x}, \omega) \cos^+ \theta'_{\mathbf{x}} \, d\omega' = L^a \, a(\mathbf{x}, \omega),$$

where

$$a(\mathbf{x}, \omega) = \int_{\Omega'} f_r(\omega', \mathbf{x}, \omega) \cos^+ \theta'_{\mathbf{x}} \, d\omega'$$

is the *albedo* of the surface. The albedo is the probability that a photon arriving at a surface from the viewing direction is not absorbed by the rough surface (it corresponds to the Fresnel function on the level of microfacets). Note that the albedo is not independent of the BRDF but can be expressed from it, and with the exception of diffuse surfaces, it is not constant but depends on the viewing direction. For diffuse surfaces, the correspondence of the albedo and the diffuse reflectivity is

$$a(\mathbf{x}) = \int_{\Omega'} k_d(\mathbf{x}) \cos^+ \theta'_{\mathbf{x}} \, d\omega' = k_d(\mathbf{x})\pi. \tag{1.12}$$

Simplified models often replace the albedo by a constant *ambient reflectivity* k_a that is set independently of the diffuse and glossy reflectivities.

Taking the Phong–Blinn model and the ambient light assumption, the local illumination approximation of the rendering equation becomes

$$L(\mathbf{x}, \mathbf{V}) = L^e(\mathbf{x}, \mathbf{V}) + k_a L^a + \sum_l L_l^{\text{in}}(\mathbf{x}) \left\{ k_d (\mathbf{N} \cdot \mathbf{L}_l)^+ + k_s \left((\mathbf{N} \cdot \mathbf{H}_l)^+ \right)^n \right\}. \tag{1.13}$$

Although we have not indicated it explicitly, the material properties k_a, k_d, k_s, and n, as well as the normal vector $\mathbf{N}$ depend on the shaded point $\mathbf{x}$. These dependences are often defined by *texture maps*. The illumination directions $\mathbf{L}_l$, viewing direction $\mathbf{V}$, and halfway vectors $\mathbf{H}_l$ depend not only on point $\mathbf{x}$ but also on the light source positions and the eye position, respectively, thus these vectors should be computed on the fly.

The local illumination formulae use only local surface properties when the reflection of the illumination of light sources toward the camera is computed. In local illumination shading, having obtained the point visible from the camera through a pixel, the reflected radiance can be determined, i.e. equation (1.13) can be evaluated without additional geometric queries. In the simplest case when even shadows are ignored, visibility is needed only from the point of the camera. To solve such visibility problems, the GPU rasterizes the scene and finds the visible points using the *z-buffer* hardware.

1.4   GLOBAL ILLUMINATION

Unlike local illumination, *global illumination methods* do not take drastic simplifications, and aim at solving the rendering equation in its original form. By accurately simulating real world physics, we expect the results to be more realistic (Figure 1.7).

Let us consider the rendering equation (1.7), which expresses the radiance of an arbitrary point $\mathbf{x}$ as a function of the radiance values of illuminating points $\mathbf{y}_1$:

$$L(\mathbf{x}, \omega) = L^e(\mathbf{x}, \omega) + \int_{\Omega'} L(\mathbf{y}_1, \omega_1') f_1 \cos^+ \theta_{\mathbf{x}}' \, d\omega_1',$$

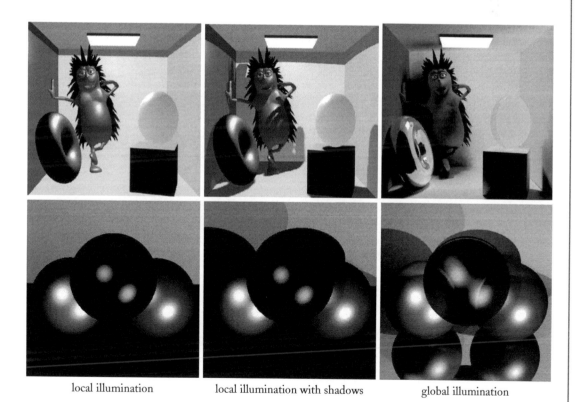

local illumination   local illumination with shadows   global illumination

FIGURE 1.7: Comparison of local illumination rendering (left), local illumination with shadows (middle), and global illumination rendering (right). Note the mirroring torus, the refractive egg, and color bleeding on the box, which cannot be reproduced by local illumination approaches.

where $f_1 = f_r(\omega_1', \mathbf{x}, \omega)$. Radiance $L(\mathbf{y}_1, \omega_1')$ at illuminating point $\mathbf{y}_1$ is not known, but we can express it using the rendering equation inserting $\mathbf{y}_1$ into $\mathbf{x}$,

$$L(\mathbf{y}_1, \omega_1') = L^e(\mathbf{y}_1, \omega_1') + \int\limits_{\Omega'} L(\mathbf{y}_2, \omega_2') f_2 \cos^+ \theta_{\mathbf{y}_1}' \, d\omega_2',$$

where $f_2 = f_r(\omega_2', \mathbf{y}_1, \omega_1')$. Using this, the radiance at point $\mathbf{x}$ can be expressed by the radiance values of points that can be reached by a single reflection:

$$L(\mathbf{x}, \omega) = L^e(\mathbf{x}, \omega) + \int\limits_{\Omega'} f_1 \cos^+ \theta_{\mathbf{x}}' \left(L^e(\mathbf{y}_1, \omega_1') + \int\limits_{\Omega'} L(\mathbf{y}_2, \omega_2') f_2 \cos^+ \theta_{\mathbf{y}_1}' \, d\omega_2' \right) d\omega_1'.$$

Repeating the same step, the radiance caused by single, double, triple, etc., reflections can be obtained. This means that to consider not only single bounce but also multiple bounce light paths, the integral of the rendering equation should be recursively evaluated at illuminating points $\mathbf{y}_1, \mathbf{y}_2, \ldots$, which leads to a sequence of high-dimensional integrals:

$$L = L_1^e + \int\limits_{\Omega'} f_1 \cos^+ \theta_{\mathbf{x}}' \left(L_2^e + \int\limits_{\Omega'} f_2 \cos^+ \theta_{\mathbf{y}_1}' \left(L_3^e + \cdots \right) d\omega_2' \right) d\omega_1'. \qquad (1.14)$$

From mathematical point of view, global illumination rendering means the solution of the rendering equation, or alternatively, the evaluation of the sequence of high-dimensional integrals for the representative wavelengths. In the following section, we first discuss efficient numerical integration techniques that can approximate such high-dimensional integrals. Then an iterative approach is presented that also obtains the solution of the rendering equation.

1.5   RANDOM WALK SOLUTION OF THE RENDERING EQUATION

Numerical quadratures approximate integrals by taking discrete samples in the integration domain, evaluating the integrand at these samples, and estimating the integral by their weighted sum. A discrete sample in the expanded integral series (equation (1.14)) is a *light path* that connects the light sources to the eye via one or more scattering points. The scattering points of the light paths are on the surface in the case of opaque objects, or can even be inside of translucent objects (*subsurface scattering* [71]). As we shall see, classical quadrature rules fail to accurately approximate high-dimensional integrals, thus we shall use Monte Carlo quadrature that takes random samples. In our case a random sample is a random light path that visits the surfaces. That is why this approach is called *random walk*.

Before going into the mathematical details, it is worth examining how the nature solves this task. In reality, a 100 W electric bulb emits about 10^{42} number of photons in each second, and the nature "computes" the paths of these photons in parallel with the speed of light no matter how complex the scene is. Unfortunately, when it comes to computer simulation, we do not have 10^{42} parallel processors running with the speed of light. This means that the number of simulated light paths must be significantly reduced, from 10^{42} to millions. To execute this reduction without significant degradation of the accuracy and image quality, we should take into account the following factors:

Dense samples. Light path samples should be found in a way that they densely fill the high-dimensional space of light paths since otherwise we might miss important light transfer possibilities.

Importance sampling. The computation should focus on those light paths which transport significant power and should not waste time simulating those whose contribution is zero or negligible.

Reuse. It is worth reusing information gained when simulating previous light paths instead of restarting the search for important light paths from scratch each time.

Let us consider first the generation of dense samples and suppose that we need to approximate the $\int_0^1 f(z)\, dz$ one-variate integral. A straightforward possibility is to take M samples $z_1, \ldots, z_M$ from a *regular grid* $(z_i = (i-1)/M)$, which leads to the following *quadrature rule*:

$$\int\limits_0^1 f(z)\, dz \approx \frac{1}{M} \sum_{i=1}^{M} f(z_i).$$

This quadrature approximates the area below the integrand as the sum of the areas of rectangles. The area of a rectangle is $f(z_i)\Delta z = f(z_i)/M$ (Figure 1.8). The error of the approximation is the total area of the "error triangles" between the function and the rectangles. The base of an error triangle is $1/M$, the number of triangles is M, and the average height of

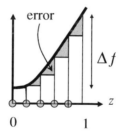

FIGURE 1.8: Error of the numerical quadrature in one dimension, taking $M = 5$ samples.

a triangle is $\Delta f / M$, where Δf is the *total variation*[3] of the integrand. Thus, the error of the quadrature is

$$\left| \int_0^1 f(z)\,\mathrm{d}z - \frac{1}{M}\sum_{i=1}^{M} f(z_i) \right| \approx \frac{\Delta f}{2M}. \qquad (1.15)$$

As we can observe, the error is inversely proportional to the number of samples, which seems to be quite reasonable.

Classical quadrature rules trace back the evaluation of higher dimensional integrals to a series of one-dimensional quadratures. Let us consider a two-dimensional integrand, $f(\mathbf{z}) = f(x, y)$:

$$\int_{[0,1]^2} f(\mathbf{z})\,\mathrm{d}\mathbf{z} = \int_0^1 \int_0^1 f(x, y)\,\mathrm{d}y\mathrm{d}x = \int_0^1 \left(\int_0^1 f(x, y)\,\mathrm{d}y \right) \mathrm{d}x = \int_0^1 F(x)\,\mathrm{d}x,$$

where $F(x)$ is the integral of the inner function. To estimate the integral of $F(x)$, we need m samples $x_1, \ldots, x_j, \ldots, x_m$ in domain x, which can be inserted into the one-dimensional quadrature. This also requires value $F(x_j)$, which itself is an integral. So for every x_j, we also need m samples y_k in domain y. The total number of the (x_j, y_k) two-dimensional samples is $M = m^2$. The two-dimensional quadrature has a form that is similar to that of the one-dimensional quadrature:

$$\int_{[0,1]^2} f(\mathbf{z})\,\mathrm{d}\mathbf{z} \approx \frac{1}{m^2}\sum_{j=1}^{m}\sum_{k=1}^{m} f(x_j, y_k) = \frac{1}{M}\sum_{i=1}^{M} f(\mathbf{z}_i).$$

Let us now estimate the error. Since the function F is estimated with a one-dimensional quadrature taking m samples, the previous result (equation (1.15)) tells us that its error is inversely proportional to m. Similarly, the integration of F requires again m samples, so the estimation error here is also inversely proportional to m. The total error of the two-dimensional integration is thus inversely proportional to $m = \sqrt{M}$. This means that the reduction of the error to its one-tenth requires the multiplication of the number of samples by a hundred, which is not appealing anymore.

Using the same reasoning, we can conclude that the error of a classical D-dimensional integral quadrature will be proportional to $M^{-1/D}$. This is unacceptable in high dimensions. For

---

[3]The total variation of a function in an interval is the absolute difference of the function at the two ends of the interval if the function is monotonous. For non-monotonous functions, the interval should be decomposed to sub-intervals where the function is monotonous, and the total variation is the sum of the total variations in these sub-intervals.

example, if the dimension is 8, the reduction of the error to its one-tenth needs 10^8 times more samples. When it comes to the solution of the rendering equation, very high, even infinitely high dimensions are possible.

The computational complexity of classical quadrature is exponential with regard to the dimension of the domain. This phenomenon is called the *dimensional explosion* or *dimensional core*. The main reason of the dimensional core is that regular grids used by classical quadrature rules do not fill the space very uniformly.

The dimensional explosion can be avoided by *Monte Carlo* [129] or *quasi-Monte Carlo* [97] integration quadratures since random points and *low-discrepancy sequences* of quasi-Monte Carlo methods are more uniform than regular grids in high-dimensional spaces. Monte Carlo methods sample light paths randomly, and their approximation error is independent of the dimension of the integration domain.

1.5.1   Monte Carlo Integration

To introduce the basics of the Monte Carlo integration, let us consider the function $f(\mathbf{z})$ that needs to be integrated in the domain V, and multiply and simultaneously divide the integrand by a *probability density* $p(\mathbf{z})$:

$$\int_V f(\mathbf{z})\, d\mathbf{z} = \int_V \frac{f(\mathbf{z})}{p(\mathbf{z})} p(\mathbf{z})\, d\mathbf{z}.$$

We can note that this integral defines the *expected value* (also called the *mean*) of random variable $f(\mathbf{z})/p(\mathbf{z})$ supposing that samples $\mathbf{z}$ are generated with the probability density $p(\mathbf{z})$:

$$\int_V \frac{f(\mathbf{z})}{p(\mathbf{z})} p(\mathbf{z})\, d\mathbf{z} = E\left[\frac{f(\mathbf{z})}{p(\mathbf{z})}\right].$$

According to the *theorems of large numbers*, an expected value can be approximated by the average of random samples $\mathbf{z}_1, \ldots, \mathbf{z}_M$:

$$E\left[\frac{f(\mathbf{z})}{p(\mathbf{z})}\right] \approx \frac{1}{M} \sum_{i=1}^{M} \frac{f(\mathbf{z}_i)}{p(\mathbf{z}_i)} = I. \tag{1.16}$$

Since the samples are random, integral estimator I is also a random variable, which fluctuates around the real quadrature value. The level of fluctuation of random variable I is

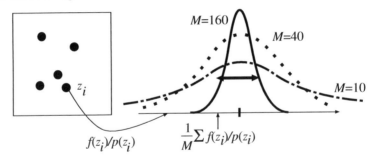

$$f(z_i)/p(z_i) \qquad \frac{1}{M}\Sigma f(z_i)/p(z_i)$$

FIGURE 1.9: The distribution of the average with respect to the number of samples. The average has Gaussian distribution and its standard deviation is inversely proportional to the square root of the number of samples.

expressed by its *variance*[4] $D^2[I]$. As the number of samples increases, the fluctuation gets smaller, which makes us more confident in the estimated value. To examine the error of this estimate, we have to realize that if $\mathbf{z}$ is a random variable of density p, then $f(\mathbf{z})/p(\mathbf{z})$ is also a random variable. Suppose that the variance of $f(\mathbf{z})/p(\mathbf{z})$ is σ^2. If the samples are independent random variables, then the variance of estimator I in equation (1.16) is

$$D^2[I] = \frac{1}{M^2}\sum_{i=1}^{M}\sigma^2 = \frac{\sigma^2}{M}. \tag{1.17}$$

Thus, the *standard deviation* (i.e. the square root of the variance) of estimator I is

$$D[I] = \frac{\sigma}{\sqrt{M}}. \tag{1.18}$$

According to the *central limit theorem*[5], estimator I will have normal distribution with standard deviation $\sigma/\sqrt{M}$ asymptotically, no matter what distribution sample $\mathbf{z}$ has. Examining the shape of the probability density of the normal distribution (Figure 1.9), we can note that the probability that the distance between the variable and the mean is less than three times the standard deviation is 0.997. Thus, with 99.7% confidence level we can say that the (probabilistic) error bound of the Monte Carlo quadrature is

$$\left| \int_{V} f(\mathbf{z})\,\mathrm{d}\mathbf{z} - I \right| < \frac{3\sigma}{\sqrt{M}}. \tag{1.19}$$

---

[4]The variance of random variable I is the expected value of the square of the difference from its mean, i.e.

$$D^2[I] = E\left[(I - E[I])^2\right].$$

[5]The central limit theorem states that the average of many independent random variables of the same distribution will have normal distribution no matter what distribution the random variables have.

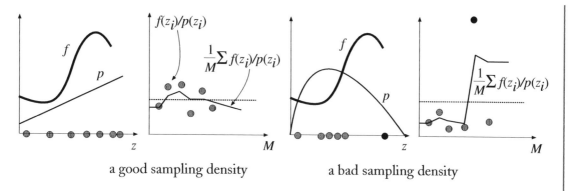

a good sampling density   a bad sampling density

FIGURE 1.10: Importance sampling.

Let us realize that this bound is independent of the dimension of the domain! This property means that by Monte Carlo quadrature the dimensional explosion can be avoided.

1.5.2   Importance Sampling

The error of Monte Carlo integration ($3\sigma/\sqrt{M}$) can be reduced either by increasing the number of samples (M) or by reducing the variance σ^2 of random variable $f(\mathbf{z})/p(\mathbf{z})$. This variance is decreased if the probability density $p(\mathbf{z})$ is more proportional to the integrand $f(\mathbf{z})$, that is, samples are concentrated where the integrand is great. This variance reduction technique is called *importance sampling*.

Figure 1.10 shows a good and a bad sampling densities. In the left of the figure, sampling density p is good, i.e. it is great where integrand f is great, thus the f/p terms in the quadrature $(1/M)\sum_{i=1}^{M} f(\mathbf{z}_i)/p(\mathbf{z}_i)$ will roughly be similar and not far from their average. As a new sample is inserted into the quadrature, the new value hardly modifies it, which means that the quadrature always remains close to the average and its fluctuation is small. In the right of the figure, the sampling density is bad, i.e. there is a region in the integration domain where integrand f is great but sampling density p is small. In this region, the small sampling density generates samples rather rarely. However, when this rare event occurs, the obtained integrand value is divided by a small probability p, which means a huge f/p term in the quadrature.

The quadrature will wander below the average for a long time until we are lucky enough to sample the important region. When it happens, the quadrature gets an enormous term, so it gets way over the average. After that it moves below the average until the next sampling of the important region. The quadrature will have a strong fluctuation, which results in a noise in the solution (Figure 1.11).

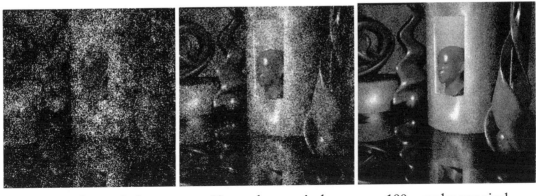

| 1 sample per pixel | 10 samples per pixel | 100 samples per pixel |

FIGURE 1.11: Noise of the Monte Carlo methods.

1.6   ITERATION SOLUTION OF THE RENDERING EQUATION

Random walk methods discussed so far look for light paths with a *depth-first search* strategy. This means that these methods extend the path by new rays until the light is connected to the eye or the further search becomes hopeless. If we apply a *breadth-first search* strategy, that is, each step extends all light paths by new rays, then we obtain a new class of solution methods. This approach is called *iteration*.

Iteration techniques realize that the solution of the rendering equation is the fixed point of the following iteration scheme:

$$L^{(m)} = L^e + \mathcal{T}_{f_r} L^{(m-1)}. \qquad (1.20)$$

Thus, if operator $\mathcal{T}_{f_r}$ is a *contraction*[6], then this scheme will converge to the solution from any initial function L_0. Since for physically plausible reflection models, the light power decreases in each reflection, the light transport operator is a contraction.

In order to store the approximating radiance functions $L^{(m)}$, usually the *finite-element method* is applied. The fundamental idea of the finite-element method is to approximate function $L(\mathbf{z})$ by a finite function series:

$$L(\mathbf{z}) \approx \sum_{j=1}^{n} L_j b_j(\mathbf{z}), \qquad (1.21)$$

---

[6]An operator or mapping is a contraction if the size of the mapped result is smaller than the size of the original. The size of a non-negative function, like the radiance, can be defined as its integral over the whole domain.

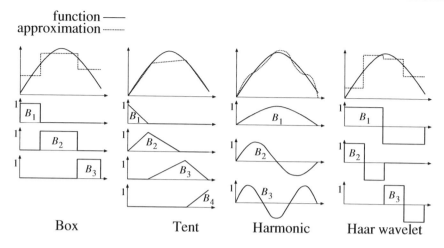

FIGURE 1.12: Basis function systems and finite-element approximations.

where $b_j(\mathbf{z})$ functions are the predefined *basis functions* and L_j values are the coefficients describing particular function $L(\mathbf{z})$.

There are many possible choices for the basis functions with different advantages and disadvantages (Figure 1.12). The particularly popular choices are the following:

Box basis functions partition the domain into subdomains and each basis function is 1 in exactly one subdomain and zero elsewhere. The coefficients are the piecewise constant approximations of the function. Box basis functions became very popular in radiosity algorithms.

Tent functions require the identification of sample points in the subdomain. Each basis function is 1 at exactly one sample point and linearly decreasing to zero between this sample point and the neighboring sample points. This corresponds to the piecewise linear approximation of the function and to *linear interpolation* which is directly supported by the graphics hardware. Sample points can be the vertices when Gouraud shading is applied, or the points corresponding to the texel centers when bi-linear texture filtering executes linear interpolation.

Harmonic functions correspond to the Fourier series approximation.

Haar wavelets also provide piecewise constant approximation, but now a basis function is nonzero in more than one subdomain.

1.7   APPLICATION OF FORMER GI RESEARCH RESULTS IN GPUGI

To develop GPU-based photo-realistic rendering algorithms, one option is to follow the research of the era when global illumination methods were running on the CPU, and try to port those algorithms onto the GPU. For example, we can revisit the research on efficient ray-shooting [109, 110, 103, 41], or we should remember that successful CPU solutions recognized that the visibility and illumination information gathered when generating a light path should be *reused* for other paths as well. *Photon mapping* [68], *instant radiosity* [78], and iterative stochastic radiosity [135] all reuse parts of the previously generated paths to speed up the computation.

However, *GPU supported global illumination* (*GPUGI*) is not just porting already known global illumination algorithms to the GPU. The GPU is a special purpose hardware, so to efficiently work with it, its special features and limitations should also be taken into account. This consideration may result in solutions that are completely different from the CPU-based methods. In this book, both the porting of CPU approaches and the development of GPU-only methods will be examined.

CHAPTER 2

Local Illumination Rendering Pipeline of GPUs

In this chapter, we review the incremental triangle rendering pipeline working with the local illumination model and discuss how its operations are supported by the graphics hardware (Figure 2.1).

The virtual world consists of objects. Every object is defined in its own *modeling space* where the definition is simple. Object definitions are mathematical formulae that identify which points belong to the surface of the object. On the other hand, incremental rendering algorithms of the GPU can process only triangles, thus the virtual world description should be translated to a set of *triangle meshes* with some level of fidelity. The approximation of surfaces by triangle meshes is called *tessellation*. Triangles output by the tessellation are defined by triplets of vertices that are also in the modeling space. The Cartesian coordinate system of modeling space is right handed in OpenGL, but Direct3D prefers a left handed system here. For every vertex, local surface characteristics required for shading are also calculated. These usually include the *normal vector* of the surface before tessellation at this point, and BRDF parameters, such as the ambient, diffuse and specular reflectivities, the shininess, or a texture address that references values of the same parameters stored in the texture memory.

The triangle list defined by vertices associated with surface characteristics is the input of the incremental rendering process. The array in graphics memory containing records of vertex data (position and local shading parameters) is called the *vertex buffer*. Which vertices constitute a triangle can be implicitly determined by the order of vertices in the vertex buffer, which is called *non-indexed drawing*. The more general solution is specifying an array of integer indices called the *index buffer*, where consecutive triplets define triangles by listing the indices of the three vertices in the vertex buffer, eliminating the need of listing vertex coordinates shared by several triangles multiple times.

In order to perform shading, vertex positions and normals have to be transformed into the coordinate system where the camera and lights are specified. This is the *world space*. Transformation parameters define where the object actually resides in the virtual world, and

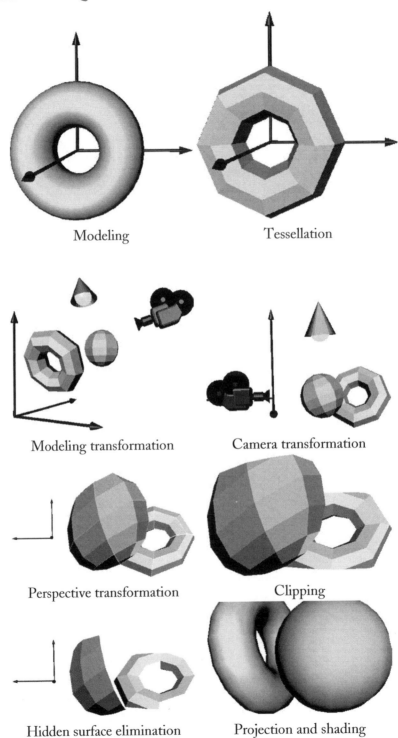

Modeling

Tessellation

Modeling transformation

Camera transformation

Perspective transformation

Clipping

Hidden surface elimination

Projection and shading

FIGURE 2.1: Steps of incremental rendering.

the computation itself is referred to as the *modeling* or *world transformation*. Making this transformation time dependent, objects can be *animated* in the virtual world.

It is not unusual that multiple instances of the same geometry have to be drawn. They might differ only at their world transformations, but variations in color or texture are also possible. Records specifying instance-specific data can be gathered in an array, which we will refer to as the *instance buffer*. It can also be seen as a secondary vertex buffer. Every vertex in the model mesh must be processed with every record of instance data, thus actual drawn vertices are defined by combinations of two records from the two buffers.

Where a triangle must be drawn on the screen is found by applying the *camera transformation*, the *perspective transformation*, and finally the *viewport transformation*. The camera transformation translates and rotates the virtual world to move the camera (also called the *eye*) to the origin of the coordinate system and to get it to look parallel with the z-axis. Perspective transformation, on the other hand, distorts the virtual world in a way that viewing rays meeting in the virtual camera become parallel to each other. This means that after perspective transformation, the more complicated *perspective projection* of the camera can be replaced by simple *parallel projection*. Perspective transformation warps the viewing pyramid to be an axis-aligned box that is defined by inequalities $-1 < x < 1$, $-1 < y < 1$, and $0 < z < 1$ (another popular choice taken by OpenGL is $-1 < z < 1$). Triangles are clipped to this box to remove those parts that fell outside of the viewing pyramid. To emphasize that clipping happens here, the space after perspective transformation is called *clipping space*, which can also be considered as *normalized screen space*.

Taking into account the resolution and the position of the *viewport* on the screen, a final transformation step, called *viewport transformation* scales and translates triangles to *screen space*. The unit of screen space is the pixel, and the X, Y coordinates of a point directly identify that pixel onto which this point is projected. In screen space the *projection* onto the 2D camera plane is trivial, only the X, Y coordinates should be kept from the X, Y, Z triplet. The Z coordinate of the point is in $[0, 1]$, and is called the *depth value*. The Z coordinate is used by *visibility computation* since it decides which point is closer to the virtual camera if two points are projected onto the same pixel. We shall use capital letters for the screen space coordinates in this book, to emphasize their special role.

In screen space, every projected triangle is rasterized to a set of pixels. When an internal pixel is filled, its properties, including the depth value and shading data, are computed via incremental linear interpolation from the vertex data. For every pixel, a shading color is computed from the interpolated data. Besides the *color buffer memory* (also called *frame buffer*), we maintain a *depth buffer* (also called *z-buffer* or *depth stencil texture*), containing screen space depth, that is the Z coordinate of the point whose color value is in the color buffer. Whenever a triangle is rasterized to a pixel, the color and the depth are overwritten only if the new depth value is less than the depth stored in the depth buffer, meaning the new triangle fragment is closer to the viewer. As a result, we get a rendering of triangles correctly occluding each other in 3D. This process is commonly called the *depth buffer algorithm*. The depth buffer algorithm is also an example of a more general operation, which computes the pixel data as some function of the new data and the data already stored at the same location. This general operation is called *merging*.

Pixel colors written into the frame buffer will be read regularly by the *display refresh* hardware and are used to modulate the intensities of red, green, and blue points on the display surface of the monitor.

2.1   EVOLUTION OF THE FIXED-FUNCTION RENDERING PIPELINE

A graphics card having the vertex buffer in its input and the color buffer at its output, and supporting *transformation*, *lighting*, *rasterization*, *texturing*, and *merging*, implements the complete process of rendering a triangle mesh, and all tasks included in the standard 3D libraries could be executed on the GPU (Figure 2.2).

Current, highly programmable graphics hardware evolved from simple monitor adapters. The task of these monitor adapters was barely more than to store an image in the *frame buffer memory*, and channel the data to control the electron beam lighting monitor pixels. Note that this was only the last stage of the incremental rendering pipeline. As rendering was expected to deliver higher performance, hardware support was added to execute the other stages as well.

During the discussion of the elements of the fixed-function pipeline, we follow a chronological order, and visit the hardware states according to their date of introduction.

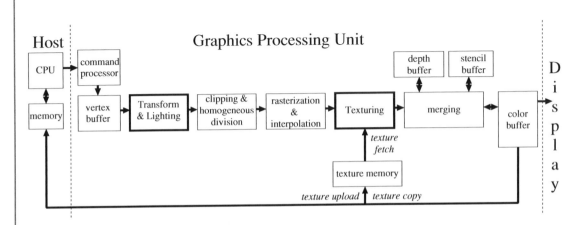

FIGURE 2.2: Fixed-function pipeline.

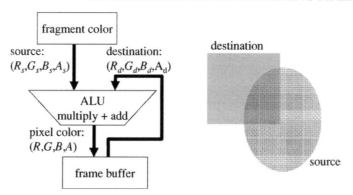

FIGURE 2.3: Blending unit that computes the new pixel color of the frame buffer as a function of its old color (destination) and the new fragment color (source).

2.1.1   Raster Operations

The frame buffer memory was soon equipped with simple arithmetic–logic functionality, which computed some function of the old pixel color and the new pixel color, and stored the result. This feature could implement *raster operations* and *blending* needed for *transparent surface rendering* (Figure 2.3). Extending the frame buffer with non-displayable data, such as with the *depth buffer* or *stencil buffer*, more sophisticated *merging* (also called *compositing*) operators could also be realized. The comparison of the new and old depth values could trigger the update of the depth and color buffers, allowing the implementation of the *depth buffer algorithm*.

Stencil buffer bits are used as flags set when a pixel is rendered to. In a consecutive rendering pass, the *stencil test* can be used to discard pixels previously not flagged. This way, algorithms like planar mirror reflections or shadows can be supported.

Merging may change or even ignore the supplied pixel color, depending on the associated depth and the stored depth, color, and stencil values. To emphasize that the supplied color and the depth might be changed or ignored, these are called *fragment* data. The result of the merging, on the other hand, is called the *pixel color* (Figure 2.3).

2.1.2   Rasterization With Linear Interpolation

Raster adapters began their real revolution when they started supporting incremental 3D graphics, earning them the name graphics accelerators. Indeed, the first achievement was to implement *rasterization* and *linear interpolation* in hardware very effectively, by obtaining values in consecutive pixels using a single addition.

Rasterization works in *screen space* where the X, Y coordinates of the vertices are equal to that pixel coordinates where these vertices are projected. The vertices may have additional properties, such as a Z coordinate in screen space and the R, G, B color values. When a triangle

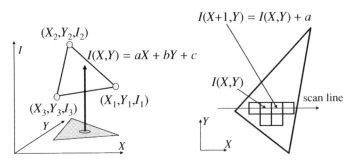

FIGURE 2.4: Incremental linear interpolation of property I in screen space.

is rasterized, all those pixels are identified which fall into the interior of the projection of the triangle by a scan line triangle filling algorithm. A *scan line* is a horizontal line that corresponds to a single row of pixels. For each scan line, the intersections with the triangle edges and the scan line are determined, and pixels between these intersections are visited from left to right incrementing the X pixel coordinate. The properties of the individual pixels are obtained from the vertex properties using linear interpolation.

Let us discuss the details of linear interpolation by taking a single *property* I that can stand for the Z coordinate (depth) or for any of the color channels. For the three vertices, the pixel coordinates and this property are denoted by (X_1, Y_1, I_1), (X_2, Y_2, I_2), and (X_3, Y_3, I_3), respectively. So we assume that I is a linear function of the pixel coordinates, that is

$$I(X, Y) = a X + b Y + c,$$

where a, b, and c are constants of the triangle. The evaluation of the function $I(X, Y)$ can take advantage of the value of the previous pixel:

$$I(X + 1, Y) = I(X, Y) + a,$$

which means that the property of the pixel can be calculated by a single addition from the property of the previous pixel.

Constant a can be obtained from the constraint that substituting the vertex coordinates into X, Y we should get back the vertex property, that is

$$I_1 = a X_1 + b Y_1 + c,$$
$$I_2 = a X_2 + b Y_2 + c,$$
$$I_3 = a X_3 + b Y_3 + c.$$

Solving this equation system for a, we obtain

$$a = \frac{(I_2 - I_1)(Y_3 - Y_1) - (I_3 - I_1)(Y_2 - Y_1)}{(X_2 - X_1)(Y_3 - Y_1) - (X_3 - X_1)(Y_2 - Y_1)}. \qquad (2.1)$$

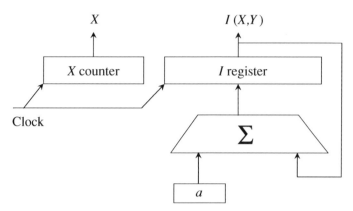

FIGURE 2.5: The hardware interpolating a single property I consists of an adder ($\sum$) that increments the content of a register by a in each clock cycle. The register should store non-integers. Pixel address X is incremented by a counter to visit the pixels of a scan line.

Since the increment a is constant for the whole triangle, it needs to be computed only once for each property. The denominator of the fraction in equation (2.1) equals the Z coordinate of the triangle normal vector in screen space, whose sign indicates whether the triangle is *front-facing* or *back-facing*. Thus, this calculation is done by a processing step called *triangle setup* that is also responsible for *back-face culling*.

The calculation of property I requires just a single addition per pixel (Figure 2.5). When it comes to hardware implementation, with as many adder units as properties we have, the new pixel values are generated in every clock cycle. The I coordinate values along the edges can also be obtained incrementally from the respective values at the previous scan line. This algorithm simultaneously identifies the pixels to be filled and computes the properties with linear interpolation.

At this stage of evolution, graphics hardware interpolated the depth coordinate and color channel values that were directly given in the vertex buffer. Thus, it was capable of executing *Gouraud shading* with the *depth buffer algorithm*.

2.1.3 Texturing

Triangle mesh models have to be very detailed to offer a realistic appearance. An ancient and essential tool to provide the missing details is *texture mapping*. Texture images are stored in the graphics card memory as 2D arrays of color records. An element of the texture image is called the *texel*. How the texture should be mapped onto triangle surfaces is specified by the texture coordinates assigned to every vertex. Thus, the vertex buffer does not only contain position and color data, but also the texture coordinates. These are linearly interpolated within triangles as described above. For every pixel, the interpolated value is used to fetch the appropriate color

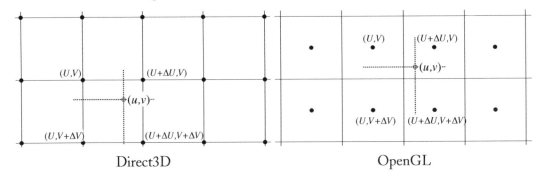

FIGURE 2.6: Texture filtering in Direct 3D and OpenGL.

from the texture memory. A number of filtering techniques combining more texel values may also be applied (Figure 2.6).

In the case of *bi-linear filtering*, the four texels closest to the interpolated texture coordinate pair (u, v) are fetched. Let these be $T(U, V)$, $T(U + \Delta U, V)$, $T(U + \Delta U, V + \Delta V)$, and $T(U, V + \Delta V)$. Here, ΔU and ΔV are the texture space distances between rows and columns of the texture image, respectively. The filtered value returned for (u, v) is then

$$T(U, V)u^* v^* + T(U + \Delta U, V)(1 - u^*)v^*$$
$$+ T(U + \Delta U, V + \Delta V)(1 - u^*)(1 - v^*) + T(U, V + \Delta V)u^*(1 - v^*),$$

where

$$u^* = \frac{u - U}{\Delta U}, \qquad v^* = \frac{v - V}{\Delta V}.$$

The value read from the texture memory can simply be used as the color of the fragment (*replace texturing mode*), but this completely ignores lighting. Lighting can be incorporated if the read value is multiplied by the color interpolated from the vertex colors (*modulative texturing mode*). If vertex colors are computed as the *irradiance* obtained with some illumination model, i.e. assuming that the surface is white, then the interpolated color can be used to modulate the original surface color fetched from the texture image.

Textures already allow for a great degree of realism in incremental 3D graphics. Not only do they provide detail, but missing shading effects, shadows, indirect lighting may be painted or pre-computed into textures. However, these static substitutes respect neither dynamic scenes nor changing lighting and viewing conditions.

2.1.4   Transformation

Rasterization algorithms assume that objects are already transformed to *screen space*. However, our virtual world is defined in other spaces. Usually each object is given in its own *modeling space*. Objects meet in the common *world space* where the camera is also specified. Thus, before rasterization is started objects need to be transformed from modeling space to world space, and according to the camera definition from world space to screen space. Supporting these transformations on the graphics card was a straightforward advancement.

The graphics hardware receives vertex positions that are given in modeling space with homogeneous coordinates. A point of the *Cartesian coordinates* (x_c, y_c, z_c) can be defined by the quadruple of *homogeneous coordinates* $[x_c w, y_c w, z_c w, w]$ using an arbitrary, nonzero scalar w. This representation owns its name to the fact that if the elements of the quadruple are multiplied by the same scalar, then the represented point will not change. From homogeneous quadruple $[x, y, z, w]$, the Cartesian coordinates of the same point can be obtained by *homogeneous division*, that is as $(x/w, y/w, z/w)$. Intuitively, homogeneous quadruple $[x, y, z, w]$ can be imagined as a point which is at the direction of another point (x, y, z) given in the Cartesian coordinates, but its distance is scaled inversely proportional to w. If $w = 1$, then the first three components of the quadruple are the Cartesian coordinates. If $w = 1/2$, for example, then quadruple $[x, y, z, 1/2]$ is equivalent to quadruple $[2x, 2y, 2z, 1]$, and we can recognize that it is a point that is in the direction of (x, y, z) from the origin of the coordinate system, but two times farther away from the origin. If $w = 0$, then the point of $[x, y, z, 0]$ is still in the direction of (x, y, z), but at "infinity". Such points cannot be expressed by the Cartesian coordinates, and are called *ideal points*.

Note that this intuition also shows that $[0, 0, 0, 0]$ cannot represent a point, and must be excluded from the valid quadruples.

Homogeneous coordinates have several advantages over the Cartesian coordinates. When homogeneous coordinates are used, even parallel lines have an intersection (an ideal point), thus the singularity of the Euclidean geometry caused by parallel lines is eliminated. Homogeneous linear transformations include *perspective projection* as well, which has an important role in rendering, but cannot be expressed as a linear function of the Cartesian coordinates. Most importantly, the widest class of transformations that preserve lines and planes are those which modify homogeneous coordinates linearly. Let us now discuss this property.

Incremental rendering needs to transform points, line segments, and triangles. Lines and planes can be defined by linear equations also in homogeneous coordinates. The *parametric equation* of a *line* defined by points $[x_1, y_1, z_1, w_1]$ and $[x_2, y_2, z_2, w_2]$ is

$$[x(t), y(t), z(t), w(t)] = [x_1, y_1, z_1, w_1]t + [x_2, y_2, z_2, w_2](1 - t),$$

where $t \in (-\infty, \infty)$. If parameter t is restricted to unit interval $[0, 1]$, we get the equation of a *line segment* of end points $[x_1, y_1, z_1, w_1]$ and $[x_2, y_2, z_2, w_2]$.

The *implicit equation* of a *plane* is

$$n_x x + n_y y + n_z z + dw = 0, \tag{2.2}$$

where (n_x, n_y, n_z) is the *normal vector* of the plane. Note that the solution of this equation does not change if all parameters n_x, n_y, n_z, d are multiplied by the same scalar. This means that the quadruple $[n_x, n_y, n_z, d]$ also has the homogeneous property. If we use a scaling that makes the normal vector (n_x, n_y, n_z) having unit length, then d expresses the signed distance of the plane from the origin in the direction of the normal vector. The equation of the plane can also be given in the matrix form:

$$[x, y, z, w] \cdot \begin{bmatrix} n_x \\ n_y \\ n_z \\ d \end{bmatrix} = 0. \tag{2.3}$$

Note that if four element row vectors represent points, then four element column vectors define planes. Points and planes are *duals*.

When a line segment is transformed, we wish to transform its two end points and connect them to form the transformed line segment. On the other hand, a triangle can also be transformed by processing only the vertices if we can make sure that the transformation maps triangles to triangles, thus the transformed vertices define the transformed object. This means that the applicable transformations must be limited to those which map line segments to line segments and triangles to triangles. The set of transformations that meet this requirement is equivalent to the linear transformations of homogeneous coordinates, that is, which can be expressed as a multiplication with a 4×4 matrix $\mathbf{T}$:

$$[x', y', z', w'] = [x, y, z, w] \cdot \mathbf{T}.$$

We can see that homogeneous linear transformations map lines to lines and planes to planes by observing that they have linear equations that are converted to also linear equations by a matrix multiplication. We shall assume that the transformation matrix is invertible, otherwise it can happen that a line degrades to a point, or a plane degrades to a line or to a point. Let us first transform the points of a line defined by its parametric equation:

$$[x(t), y(t), z(t), w(t)] \cdot \mathbf{T} = ([x_1, y_1, z_1, w_1]t + [x_2, y_2, z_2, w_2](1 - t)) \cdot \mathbf{T} =$$

$$[x'_1, y'_1, z'_1, w'_1]t + [x'_2, y'_2, z'_2, w'_2](1 - t),$$

where

$$[x_1', y_1', z_1', w_1'] = [x_1, y_1, z_1, w_1] \cdot \mathbf{T}, \qquad [x_2', y_2', z_2', w_2'] = [x_2, y_2, z_2, w_2] \cdot \mathbf{T}.$$

Note that the equation of the transformed set is similar to that of the line, thus it is also a line.

The discussion of the plane transformation is based on the recognition that if we transform the already transformed points $[x', y', z', w'] = [x, y, z, w] \cdot \mathbf{T}$ with the inverse of $\mathbf{T}$, then we get the original plane back, which satisfies the plane equation:

$$\left([x', y', z', w'] \cdot \mathbf{T}^{-1} \right) \cdot \begin{bmatrix} n_x \\ n_y \\ n_z \\ d \end{bmatrix} = 0.$$

Exploiting the associativity of matrix multiplication (i.e. the possibility of rearranging the parentheses) and introducing the transformed plane quadruple as

$$\begin{bmatrix} n_x' \\ n_y' \\ n_z' \\ d' \end{bmatrix} = \mathbf{T}^{-1} \cdot \begin{bmatrix} n_x \\ n_y \\ n_z \\ d \end{bmatrix}, \tag{2.4}$$

we can see that transformed points also satisfy a plane equation

$$[x', y', z', w'] \cdot \begin{bmatrix} n_x' \\ n_y' \\ n_z' \\ d' \end{bmatrix} = 0.$$

Thus, planes are transformed to planes. An important consequence of this proof is that the quadruple of a plane should be transformed with the inverse of the transformation matrix. If the quadruple of the plane parameters, including the normal vector, are considered not as a column vector but as a row vector similarly to the quadruple of points, then the inverse transpose of the transformation matrix should be used.

From the point of view of GPU hardware, the transformations to camera space and then from the camera to normalized screen space (or clipping space) are given as 4×4 homogeneous linear transformation matrices, the world-view, and the perspective one, respectively. In order for a hardware to support transformations, 4×4 matrix and four-element vector multiplication units need to be included that process the homogeneous vertex coordinates. The transformation matrices should be applied for all vertices, and are not stored in the vertex buffer, but in a few registers of the graphics card, which are allocated for such *uniform* parameters. Whenever these matrices change due to object or camera animation, the vertex buffer does not need to be altered.

2.1.5   Per-vertex Lighting

Together with transformation support, GPUs also took the responsibility of illumination computation. To allow the GPU to compute lighting, instead of including color values in the vertex buffer, the data required to evaluate the local shading formula are stored. This includes the surface normal and the diffuse and glossy reflection coefficients. Light positions, directions, and intensities are specified as uniform parameters over all the vertices.

The color of the vertex is computed by the following *illumination formula*:

$$L^e + k_a L^a + \sum_l L_l^{in}(\mathbf{x}) \left\{ k_d(\mathbf{N} \cdot \mathbf{L}_l)^+ + k_s \left((\mathbf{N} \cdot \mathbf{H}_l)^+\right)^n \right\}, \qquad (2.5)$$

where L^e is the own emission, L^a is the ambient light intensity, l runs for a limited number of abstract light sources, $L_l^{in}(\mathbf{x})$ is the light source intensity that might be scaled according to the distance for point sources, k_d, k_s, and k_a are the spectral diffuse, glossy, and ambient reflectivities, respectively, and n is the shininess of the material. The normal vector $\mathbf{N}$ is transformed to camera space with the inverse transpose of the model-view transformation (equation (2.4)). Camera space light source direction or location is obtained by the graphics API. In camera space the camera position is the origin. Using these, the unit illumination direction $\mathbf{L}$, viewing direction $\mathbf{V}$, and halfway direction $\mathbf{H}$ can be obtained for every vertex by the GPU. Since light source and material properties depend on the wavelengths, at least three such formulae should be evaluated for the wavelengths of red, green, and blue.

Having computed vertex colors, the transformation, the rasterization, and the linear interpolation are performed to color the pixels, just like without lighting.

2.2   PROGRAMMABLE GPUS

Although the *fixed-function pipeline* provided a particular solution to incremental rendering, it greatly constrained how its immense parallel computing power could be leveraged. It was not flexible, and techniques that did not directly fit into its basic concept could not be supported. Billboards and skeleton-animated characters, for instance, require different or more complex transformations than rigid objects. Lighting using more realistic BRDF models than the Phong–Blinn formula was not possible. Computing pixel colors as an interpolated color modulated by a single texture was also crippling limitation.

Alleviating the problems of the fixed-function pipeline, and turning the GPU into a general purpose stream processor, the former transformation and lighting stage and the texturing stage were replaced by programmable *vertex shader* and *fragment shader* stages (Figure 2.7).

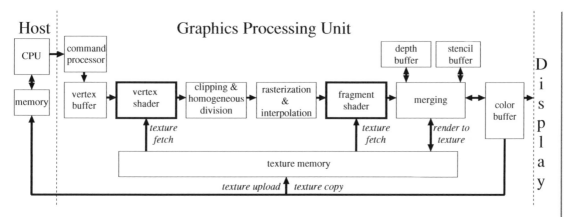

FIGURE 2.7: Shader Model 3 architecture with vertex and fragment shaders, render-to-texture, and vertex textures.

The interface between them has also been extended: not only the vertex color, but a number of arbitrary vertex shader output properties could be interpolated and passed to the fragment shader. One of the most straightforward consequences was that by interpolating shading normals, Gouraud shading could be replaced by *Phong shading*, moving lighting to the fragment shader. This is usually referred to as *per-pixel lighting* (Figure 2.8).

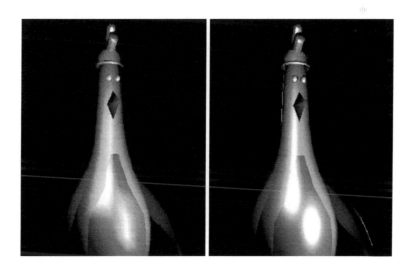

FIGURE 2.8: A chicken with per-vertex (left) and per-pixel lighting (right). Note that per-vertex lighting and linear color interpolation smears specular highlights incorrectly. This problem is solved by per-pixel lighting.

2.2.1   Render-to-texture and Multiple Render Targets

In the classic case, fragments colored by the fragment shader are finally written into the frame buffer. Textures needed in fragment processing are usually uploaded by the CPU. To make more use of a computed image, the early solution was to copy the content of the frame buffer to the texture memory, and thus use previous images later as textures. This way some dependency is introduced between consecutive rendering passes. This feature has become crucial in many applications, so its limitations needed to be addressed. The frame buffer allows just 8-bit data to be stored, and it is rather slow to copy it to the texture memory. Both problems are eliminated by the *render-to-texture* feature of advanced graphics cards, which allows the final output to be redirected from the frame buffer to a given texture. With this feature some passes may perform general purpose or global illumination computations, write the results to textures, which can be accessed later in a final gathering step, rendering to the frame buffer.

The frame buffer or the texture memory, which is written by the fragment shader and the merging unit, is called the *render target*. As output colors are no longer written to a frame buffer directly associated with the display device, it is also possible to output values simultaneously to several textures. However, render targets still have to have the same dimensions, and a fragment shader will write values to all targets at the same pixel location. This feature is called *multiple render targets*.

2.2.2   The Geometry Shader

While vertex and fragment shaders have evolved from small, specialized units to general stream processors, they have kept one record of the output for every record of the input scheme. It is possible to have a primitive discarded by back-face culling and clipping, and a fragment discarded by the depth and stencil tests, but these can only achieve some empty pixels in the render target, not variable-length output. The solution is provided by the *geometry shader* stage of Shader Model 4.0 GPUs. It works on vertex shader output records (processed vertices), and outputs a varying number of similar records. This data can be directed into an *output stream*, which is, structurally, a vertex buffer.

There are two major implications. Most relevant to GPUGI is that we can even skip rasterization, fragment shading, and frame buffer operations, and only output a variable length array of data. It can open the door to the GPU implementation of random-walk-based GI algorithms. However, this is yet to be implemented and proven to be effective. In this book, most of the introduced techniques are capable of running on the Shader Model 3.0 graphics hardware, and do not utilize geometry shaders.

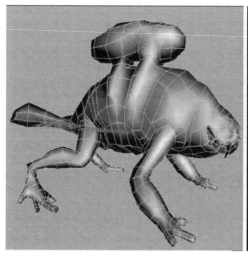 

Low-poly model   Subdivided mesh with displacement mapping

FIGURE 2.9: A typical geometry shader application: on the fly Catmull–Clark subdivision of Bay Raitt's lizard creature [3]. Normal mapping is added by the fragment shader.

The other, more advertised feature of geometry shaders is that they can move geometry processing operations like tessellation to the GPU. Procedural geometries, adaptive subdivision [3] (Figure 2.9), displacement mapping with correct silhouettes [66, 143] (Figure 2.10), or the generation of shadow volumes are prime examples.

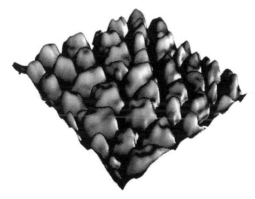

FIGURE 2.10: Another typical geometry shader application: displacement mapping with correct silhouettes obtained by local ray tracing at 200 FPS on an NV8800 GPU.

2.3 ARCHITECTURE OF PROGRAMMABLE GPUS

Throughout this book, we will adopt the latest, most general theoretic view of graphics hardware, which is that of Shader Model 4.0, or Direct3D 10. However, examples and the implementations of global illumination algorithms will conform to the Shader Model 3.0 standard (Figure 2.7), and do not require more advanced hardware to run.

Figure 2.11 shows the block model of the Shader Model 4.0 GPU architecture. The blocks belong to three main categories: resources, programmable and fixed-function pipeline stages, and render states.

2.3.1 Resources

Resources are blocks of data residing in the graphics memory. These are read from and written to by different *pipeline stages*, and they are also accessible from the CPU through the *PCI Express* interface. Resources can mostly be thought of as untyped, continuous memory segments. What was an output buffer in a pass could easily be re-interpreted as an input buffer in the next. Vertex, instance, output and constant buffers are arrays of data records. *Index buffers* must be arrays of integers. *Textures* are 1D, 2D, 3D arrays of 1, 2, 3, or 4-channel, integer or floating point color records, or six-element collections of 2D cube-face textures.

The *depth buffer* used in hidden surface elimination is called *depth stencil texture* in the Direct3D terminology. A render target or a depth stencil texture must always be a 2D subset, but multiple render targets are also allowed. *Constant buffers* and textures are accessible by all shader stages. They are both random access arrays, the difference being that textures can be addressed by floating point coordinates and filtered, while constant buffers are faster to access but limited in size. Resources, particularly if they are large, should rarely be updated from the CPU program. They are either loaded only once, or generated on the GPU itself, as render targets or output buffers.

2.3.2 Pipeline Stages and Their Control by Render States

How various pipeline stages of the GPU pipeline operate is governed by an extensive set of *render states*. For programmable pipeline stages, a program must be specified, which is typically written in the Cg or HLSL language, compiled at runtime, and uploaded to the graphics card. Furthermore, *uniform parameters* can also be set, meaning shader input data identical for all vertices, primitives or fragments. As opposed to constant buffers, uniform parameters are typically regularly set from the CPU.

The operation of the pipeline is triggered by issuing a *draw call* by the CPU. After a *draw call*, the *input assembler* stage will construct the vertex data for the vertex shader from the vertex and instance buffers. All vertices for all instances have to be processed. How elements in

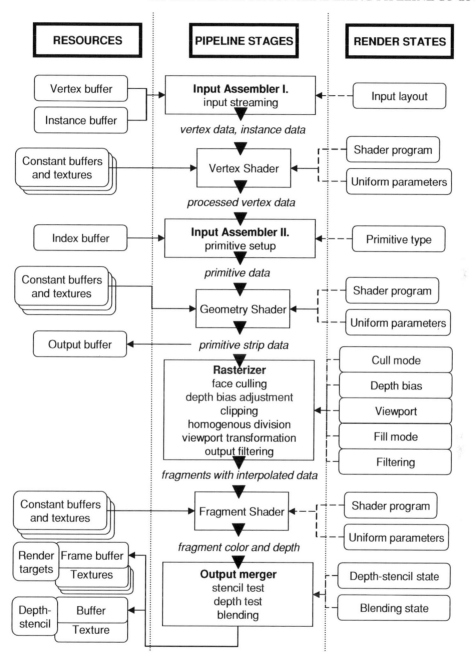

FIGURE 2.11: The pipeline model of Shader Model 4.0 GPUs.

the input records are mapped to vertex shader parameters is defined by the render state *input layout*. In the shader program, every parameter must have a *semantic* identifier specified. The input layout is a list mapping semantic identifiers to vertex buffer record fields. It is also known as a *vertex format description*.

The *vertex shader* receives a record containing both data from the vertex buffer and the instance buffer (a vertex of an instance). In the fixed-function pipeline, this stage would be responsible for transforming the vertex to homogeneous *clipping space* and may modify the color properties if lighting computations are also requested. A vertex shader program, on the other hand, can output an arbitrary vertex data record, as long as it includes normalized screen space position in homogeneous coordinates for the clipping, depth testing and rasterization operations.

The next stage, still regarded to be the part of the input assembler, combines processed vertices belonging to the same primitive. How primitives are assembled from the vertices is governed by the *primitive type* render state. In Direct3D 9, this is given as a parameter to the draw call. In OpenGL, on the other hand, it is handled by the command processor.

The geometry shader stage receives vertex records along with primitive and *adjacency* information. It may just pass them on as in the fixed-function pipeline, or spawn new vertices. These may be written to an output buffer, which can be used as an input vertex buffer in a consecutive pass.

The primitives generated by the geometry shader are passed on to the *rasterizer* stage, which performs the bulk of the non-programmable functionality within the pipeline. According to the *cull mode* render state, front-facing or back-facing triangles can be discarded. Facing information will also be passed down to the output merger stage, to be able to apply different operations on front-facing and back-facing triangles. By setting *depth bias* render states, some automatic adjustment can be made to avoid ambiguity when comparing depth values. The *clipping* hardware keeps only those parts where the $[x, y, z, w]$ homogeneous coordinates meet the following requirements defining an axis aligned cube of corners $(-1, -1, 0)$ and $(1, 1, 1)$ in normalized screen space, defined by the following inequalities using homogeneous coordinates:

$$-w \leq x \leq w, \; -w \leq y \leq w, \; 0 \leq z \leq w.$$

These equations are valid in Direct3D. OpenGL, however, modifies the last pair of inequalities to $-w \leq z \leq w$.

Clipping may introduce new vertices for which all properties (e.g. texture coordinates or color) are linearly interpolated from the original vertex properties. After clipping the pipeline executes *homogeneous division*, that is, it converts the homogeneous coordinates to the Cartesian ones by dividing the first three homogeneous coordinates by the fourth (w). The points are then transformed to *viewport space* where the first two Cartesian coordinates select the pixel in which this point is visible. Finally, the rasterization hardware fills the projection of the primitive on the X, Y plane, visiting each pixel that is inside the projection (or only the boundaries, depending on the *fill mode* render state). During filling, the hardware interpolates all vertex properties to obtain the attributes of a particular fragment.

Because of finite pixels, the result image is going to be a discretely sampled version of the intended continuous one. Therefore, *aliasing* will manifest as visible pixel edges, disturbing to the human eye, especially during animations. Pre- or postfiltering is necessary to reduce this artifact. *Prefiltering* means that we render objects that have only smooth transitions. These are called *smooth lines* or *smooth polygons*. OpenGL has supported them for long, but hardware implementations are just emerging, with smooth lines appearing in Direct3D. *Postfiltering* means applying a higher sampling rate, and then filtering the discrete samples down to a lower resolution. This is known as *multisampling*, and is governed by render states associated with the rasterizer stage.

Fragments to be filled by the rasterizer stage are fed to the *fragment shader*. It takes the attributes of a particular fragment and computes the fragment color for every render target. It may also output a depth value for the depth test, overwriting the hardware-interpolated one. In the fixed-function pipeline with modulative texturing enabled, this stage merely performed a texture lookup and multiplied the result with the color attribute. With programmability, fragment shaders range from per-pixel lighting implementations through ray-tracing to non-graphics general purpose computations.

When final fragment colors are computed, they are not directly written to the render targets, but the *output merger* stage is responsible for the composition. First, the stencil test against the stencil buffer and the depth test against the depth buffer are performed. Note that these operations might be moved before the execution of the fragment shader if it does not output a depth value. This *early z-testing* might improve performance by not processing irrelevant fragments.

Finally, for every render target, the output merger blends the new fragment color with the existing pixel color, and outputs the result. The render target can be used as an input resource to later passes, or presented on the device screen.

CHAPTER 3

Programming and Controlling GPUs

In this chapter, we are going to give a short review of the techniques necessary to control what the GPU pipeline does, accompanied by simple programming examples. A reader deeply familiar with HLSL programming may wish to skip this introduction, and continue reading at Section 3.3.1, where we discuss how global illumination algorithms can make use of GPU programmability. However, this chapter defines concepts which serve as a foundation for more complex techniques, and reviewing the programming patterns introduced here greatly helps understanding later code examples in the book. On the other hand, the chapter is not meant to be a complete programming reference: you will also need to consult API specifications and detailed reference guides on Direct3D or HLSL programming if you engage in implementing GPUGI techniques yourself.

In order to control pipeline operation, we need to set *render states* for fixed function stages, and assign programs to programmable stages. These are done in a classic CPU program. Therefore, having a Shader Model 4.0 GPU in our hands, we should write four programs, one for each of the CPU, vertex shader, geometry shader, and the fragment shader. The CPU program uploads the shader programs to the GPU, controls the operation of the non-programmable GPU stages via render states, and feeds the pipeline with data representing the scene to be rendered. Shader programs get data sent down the pipeline, but the CPU can also set global GPU variables that can be accessed by the shaders. All these tasks are executed by calling functions of the graphics API, i.e. Direct3D or OpenGL. The shader programs are usually written in special languages, for example in *HLSL* or in *Cg*. Just like with other programming languages, HLSL programs must be compiled before they can be run on hardware. The CPU program reads shader programs—either source code or compiled—from files, compiles them if needed, and uploads them to the GPU's shader units, again using appropriate graphics API calls.

This chapter introduces HLSL, a language for shader programming, and discusses how programs written in it can be executed on the GPU using Direct3D.

3.1 INTRODUCTION TO HLSL PROGRAMMING

The *High Level Shader Language* (HLSL) is a stream programming language developed by Microsoft for use with *Direct3D*. It is very similar to the NVIDIA *Cg* shading language. In this book, we use HLSL when the vertex, geometry, and fragment shader programs are presented. However, the *Cg* implementations would be very similar or even identical in most of the cases. Due to the rapid development of GPU features, HLSL is not a static language, but progresses continuously, introducing new language constructs and eliminating previous limitations. In particular, we may talk about Shader Model 1, 2, 3, and currently 4 compatible program code. In order to tell the compiler what version of HLSL should be used, we should specify the *target profile* as one of the *compiler options*. Depending on how we perform compilation, these options are specified as parameters to an API call, command line options to a stand-alone compiler, or in an *effect file*. An HLSL effect file conveniently encapsulates render states, resource and register mappings, and shader source code with compiler options. We will describe them in more detail in Section 3.2.2.

HLSL programs for different target Shader Models are mostly backward compatible, or at least there is a compiler option to allow earlier syntax and semantics when compiling for a more advanced model. Attempting to compile later code for an earlier target profile will result in compilation errors. On the other hand, you can only run the compiled program on GPUs that support the target profile. Thus, the selection of the Shader Model limits the set of capabilities we can use and defines the hardware requirements. In this book, we give examples in Shader Model 3 compatible HLSL. In special cases, we also use Shader Model 4 compatible HLSL.

HLSL is based on the C language, but vectors of up to four elements and matrices up to 4×4 were also added to the basic types. For example, type `float4` defines a four-element vector where each element is a 32-bit `float` value. Type `float4x4` defines a 4×4 matrix. The four elements of a `float4` variable can be selected with syntax `.x`, `.y`, `.z`, `.w`, respectively, or with syntax `.r`, `.g`, `.b`, `.a`, respectively. We can even select multiple elements at once. For example, the two element vector containing the coordinates `.x` and `.w` can be selected using syntax `.xw`. This is called the *swizzle operator*. It is important to note that these are not only language constructs, but the GPUs have *Arithmetic Logic Units* (*ALU*) that can process four floats in parallel.

From the point of view of shader programming, the basic unit of pipeline operation is the *pass*. During a pass, the CPU may change uniform parameters before draw calls and trigger rendering of different geometries, but shaders and usually also render states remain constant. Rendering an image may require the use of several different shaders or render states, and thus, multiple passes. Consecutive passes rendering to the same render target constitute a *technique*.

A HLSL program specifies how the vertex, geometry, or fragment shader operates on streams of data. A shader is defined by a top-level function, which is executed independently for every element of the stream.

A HLSL program has two kinds of inputs. *Varying inputs* constitute a single record of the stream of input data. *Uniform inputs* are the global data that affect the transformation of all stream elements, and do not change with each one. Some inputs, that were considered to be part of the render state using the fixed-function pipeline, are now handled using uniform parameters. Most notably, `World`, `View`, and `Projection` transformation matrices are passed as any other uniform parameter.

A shader program is best imagined as a "transformation" that converts input data to output data. For the vertex shader, the input/output data are vertex records containing data such as position, color, texture coordinates, or the surface normal. For the geometry shader, the input/output is a set of primitives. For the fragment shader, the input is interpolated vertex data for a single fragment, and the output is fragment color and depth. Varying inputs are indicated by the `in` keyword in the parameter list of the top-level function, but this keyword is only optional since this is the default type. Uniform inputs can be defined as global variables, or specified as `uniform` in the parameter list.

The output can only be varying and describes the current output record of the stream. An output element is defined by the `out` keyword if it is included in the parameter list. Alternatively, the return value of the top-level function is automatically considered to be an output.

All outputs and varying inputs must have associated semantic labels. Some semantics, like `POSITION`, indicate a value that has been or will be interpreted as the homogeneous coordinates of a vertex, and processed by fixed-function operations like clipping. In Shader Model 4.0, these semantics are prefixed by `SV_`, e.g. `SV_TARGET` replaces `COLOR`. Other semantics, like `TEXTURE4`, merely link shader outputs to shader inputs of later programmable stages. If a vertex shader has written a value to output variable with semantics `TEXTURE4`, the fragment shader will receive the interpolated value in varying input parameter with semantics `TEXTURE4`. In early shader models, semantics explicitly mapped variables to input and output registers of shader units. With *unified shader architecture*, where the same hardware unit can fill the role of a vertex, geometry or fragment shader, and with the possible number of attributes far exceeding the number of registers dedicated to fixed function tasks, binding is more flexible. However, for the sake of clarity, we will refer to the named input and output registers of shader units when describing shader functionality. Any input and output parameters can be organized into `struct` records. In this case, semantics are not given in the function header, but in the `struct` definition.

In the following subsections we take the standard, fixed-function Direct3D operation as an example, and show how it could be implemented by programmable shaders.

3.1.1 Vertex Shader Programming

A vertex shader is connected to its input and output register sets defining the attributes of the vertex before and after the operation. All registers are of type `float4` or `int4`, i.e. are capable of storing a four-element vector. The input register set describes the vertex position (`POSITION`), colors (`COLOR0`, `COLOR1`), normal vector (`NORMAL`), texture coordinates (`TEXCOORD0`, ..., `TEXCOORD7`), etc. These many texture registers can be used in multi-texturing, or can rather be considered as registers for general purpose data. The vertex shader unit computes the values of the output registers from the content of the input registers. During this computation, it may also use global variables or *uniform parameters*. The vertex shader is a *top-level function*, not invoked by HLSL functions but indirectly from the CPU program. This function may call other functions and may declare local variables.

The following example shader realizes the vertex processing of the fixed-function pipeline when lighting is disabled. It applies the `WorldViewProj` transformations to transform the vertex from modeling to world space, from world to camera space, and from camera to homogeneous clipping space, respectively:

```
// Homogenous linear transformation from modeling to homogeneous clipping space
float4x4 WorldViewProj;

void StandardNoLightingVS(
    in  float4 Pos    : POSITION,    // Modeling space vertex position
    in  float3 Color  : COLOR0,      // Input vertex color
    in  float2 Tex    : TEXCOORD0,   // Input texture coordinates
    out float4 hPos   : POSITION,    // Clipping space vertex position
    out float3 oColor : COLOR0,      // Output vertex color
    out float2 oTex   : TEXCOORD0    // Output texture coordinates
) {
    hPos = mul(Pos, WorldViewProj);  // Transform to clipping space
    oColor = Color;                  // Copy input color
    oTex = Tex;                      // Copy texture coords
}
```

The transformation of the vertex position (given in input register `POSITION` since Pos is bound to `POSITION` using semantics declaration `Pos : POSITION`) uses the `mul` intrinsic function that multiplies the four-element vector of the vertex position by the 4×4 `WorldViewProj` matrix that transforms from modeling space to clipping space. The transformed position is written into output register `POSITION` since variable `hPos` is bound to the output `POSITION` register by declaration `out float4 hPos : POSITION`. The input color and the input texture coordinates are simply copied to the output color and output texture coordinates, respectively. Note that the colors are defined by type `float3` although the respective registers are of type `float4`.

This means that only the first three elements are used and the compiler checks whether or not the programmer mistakenly refers to the fourth element, which would cause an error. Similarly, texture coordinates are just two-element vectors in this particular example. We note that declaring a variable as float, float2, float3 when not all four elements are needed is a good programming practice, but it usually does not decrease storage space since the registers have four elements anyway, and it does not improve performance since shader ALUs are capable to execute the same operation in parallel on all four elements (in this respect, GPU shader processors are similar to the *SSE* and *3DNow!* features of modern CPUs).

The second vertex shader example executes local illumination computation for the vertices (*per-vertex lighting*), and replaces the color attribute with the result. The illumination is evaluated in camera space where the camera is at the origin and looks at the z direction assuming Direct3D (the camera would look at the $-z$ direction in OpenGL). In order to evaluate the *Phong–Blinn reflection model* (equation (2.5)), the normal, lighting, and viewing directions should be obtained in camera space. We consider just a single point light source in the example.

```
float4x4 WorldViewProj; // From modeling to homogeneous clipping
space float4x4 WorldView;      // From modeling to camera space
float4x4 WorldViewIT;    // Inverse-transpose of WorldView to transform normals

// Light source properties
float3 LightPos;      // Light position in camera space
float3 Iamb;    // Intensity of ambient light
float3 Ilight;    // Intensity of point light source

// Material properties
float3 ka, kd, ks;  // Ambient, diffuse, and glossy reflectivities
float shininess;

void StandardLightingVS(
    in   float4 Pos     : POSITION,    // Modeling space position
    in   float3 Norm    : NORMAL,      // Modeling space normal
    in   float2 Tex     : TEXCOORD0,   // Input texture coords
    out  float4 hPos    : POSITION,    // Clipping space position
    out  float3 oColor  : COLOR0,      // Output vertex color
    out  float2 oTex    : TEXCOORD0    // Output texture coords
) {
    hPos = mul(Pos, WorldViewProj); // Transform to clipping space
    oTex = Tex;                     // Copy texture coords
    // Transform normal to camera space
    float3 N = mul(float4(Norm, 0), WorldViewIT).xyz;
    N = normalize(N);
```

```
    // Transform vertex to camera space and obtain the light direction
    float3 cPos = mul(Pos, WorldView).xyz;
    float3 L = normalize(LightPos - cPos);
    // Obtain view direction (the eye at the origin in camera space)
    float3 V = normalize(-cPos);    // Viewing direction
    float3 H = normalize(L + V);    // Halfway vector
    // Evaluate the Phong-Blinn reflection
    float costheta = sat(dot(N, L));
    float cosdelta = sat(dot(N, H));
    oColor = Iamb * ka + Ilight * (kd * costheta + ks * pow(cosdelta, shininess));
}
```

This program transforms the input position to clipping space and copies the texture coordinates just like the previous vertex shader executing no lighting computations. To prepare the computation of the illumination formula, the normal vector is transformed to camera space as described in Section 2.1.4. Besides the transformation matrix that takes a point from modeling space to camera space (WorldView), we also passed the inverse transpose of this matrix (WorldViewIT) as a global, uniform variable. The normal—augmented with a zero to get a homogeneous direction vector—is multiplied with this matrix using the mul HLSL function. The .xyz swizzle operator selects the first three elements of the result. The normal vector is *normalized*, i.e. it is scaled to have unit length, with the normalize HLSL function.

Multiplying the modeling space position (Pos) and the matrix transforming to camera space (WorldView), the position in camera space can be determined and stored in the local variable cPos. Note that we used the swizzle operator to take the first three coordinates of a homogeneous quadruple, and assumed that this is equivalent to converting from homogeneous to the Cartesian coordinates. This is true only if the fourth homogeneous coordinate of the position is 1 in the vertex buffer, and the transformation to camera space is *affine*, i.e. its fourth column consists of values 0, 0, 0, 1. Since transforming from modeling space to world space usually consists of scaling, rotation, and translation, and transforming from world space to camera space is a translation plus a rotation, the assumption on affine transformation is valid.

Global variable LightPos stores the position of the light source in camera space. Taking the difference between this vector and the place vector of the point cPos, we obtain the direction of the illumination (L). This direction is normalized with the normalize HLSL function. According to the Lambert law of diffuse reflection, the cosine between the surface normal and the illumination direction needs to be computed. Since the normal vector N and the illumination direction (L) are unit length vectors, their *dot product* computed with the dot HLSL intrinsic is the cosine value. The cosine should be replaced by zero if it is negative. This operation is executed by saturation function sat. In fact, the saturation function truncates the input value to be in the [0, 1] interval.

In camera space the eye is at the origin, thus the viewing direction can be obtained by subtracting the position of the vertex from the origin, which is just $-$cPos. The viewing direction is also normalized to get vector V, and unit halfway vector H is computed. Applying the dot then the sat HLSL functions, the cosine of the specular part is obtained. Finally, equation (2.5) is evaluated. Since light source parameters Iamb, Ilight, material parameters ka, kd, ks, and output color oColor are all of type float3, this single line of program code simultaneously computes the color on the wavelengths of red, green, and blue.

3.1.2   Geometry Shader Programming

Since the geometry shader appears in Shader Model 4.0, the description and the example shown in this section cannot conform to Shader Model 3.0 syntax, contrary to the rest of the examples in this chapter.

The geometry shader receives the vertex data output by the vertex shader, organized into primitives. For the input parameter, the primitive type has to be specified. The shader does not output values to specific output registers, but appends records to a stream, which is specified as an inout parameter of the generic collection type PointStream, LineStream, or TriangleStream.

The following example shader takes a triangle primitive with vertices only containing a position value, and appends it to the output stream without modification.

```
[MaxVertexCount(3)]                    // Max number of output
vertices void TraingleGS(
    triangle float4 InPos[3],          // Triangle vertex positions array
    inout TriangleStream<float4> OutPosStream  // Triangle strip vertex stream
) {
    OutPosStream.Append( InPos[0] ); // Append vertex
    OutPosStream.Append( InPos[1] ); // Append vertex
    OutPosStream.Append( InPos[2] ); // Append vertex
    OutPosStream.RestartStrip();     // Finish triangle strip
}
```

3.1.3   Fragment Shader Programming

The *fragment shader* (also called *pixel shader*) receives the fragment properties of those pixels which are inside the clipped and projected triangles, and uniform parameters. The main role of the fragment shader is the computation of the fragment color.

The following example program executes modulative texturing. Taking the fragment input color and texture coordinates that are interpolated from vertex properties, the fragment shader looks up the texture memory addressed by the texture coordinates, and multiplies the read texture data with the fragment input color:

```
sampler2D textureSampler; // 2D texture sampler

void TexModPS(
    in  float2 Tex    : TEXCOORD0,  // Interpolated texture coords
    in  float4 Color  : COLOR0,     // Interpolated color
    out float4 oColor : COLOR       // Color of this pixel
) {
    oColor = tex2D(textureSampler, Tex) * Color;   // Modulative texturing
}
```

This program uses the `tex2D` *HSLS intrinsic* to fetch texture data. Depending on the settings of the *texture sampler* called `textureSampler`, this instruction returns either the nearest texel or reads multiple texels and applies filtering. A texture sampler can be imagined as the encapsulation of the texture map and its filtering, mipmapping, and clamping properties.

The output of the fragment shader is the output fragment color and optionally the output depth value. The output fragment color continues its trip in the rendering pipeline and enters the merging stage that may apply the depth and stencil testing, or alpha blending, and writes the resulting color to the *render target*. For a fixed-function-imitating fragment shader, the render target will typically be the frame buffer.

As mentioned, output variables can be declared as the return value of a shader program. Thus, the previous program can also be written as follows:

```
sampler2D textureSampler; // 2D texture sampler

float4 TexModPS(
    in float2 Tex    : TEXCOORD0, // Interpolated texture coords
    in float4 Color  : COLOR0     // Interpolated color
) : COLOR // Output is the COLOR register that has type float4
{
    return tex2D(textureSampler, Tex) * Color;
}
```

3.2   CONTROLLING THE GPU FROM THE CPU

To define and initiate the operation of the pipeline, the CPU program must perform the following:

1. create and bind the input vertex, instance, and index buffers;

2. create and bind render targets, the depth-stencil buffer and/or the output buffer;

3. create and bind constant buffers and textures;

4. set render states;

5. create and bind shader programs;

6. set uniform variables;

7. issue a draw call to start the pipeline.

   In OpenGL, the assembly of the input geometry through `glBegin`, `glVertex`, `glEnd`, and related calls is also part of the graphics API. These commands are passed to the *command processor*, which fills up the vertex buffer with modeling space vertices and their attributes, and controls other aspects of the pipeline as well. While this is a highly intuitive and educational method of building geometry, it is not effective and conceptually obsolete compared to mapping vertex buffers to system memory arrays. Binding resources and starting the pipeline is also a task of the command processor, usually through OpenGL extensions.

   In Direct3D, a CPU *draw call* triggers the operation of the pipeline. There might be several versions and parameterizations to the call in the graphics library API to select indexed or non-indexed drawing, control instancing, or specify working ranges within input vertex, instance, and index buffers.

   In the following subsections, we detail the CPU program assuming Direct3D API.

3.2.1   Controlling the Direct3D 9 Pipeline

An empty Direct3D 9 project (included in the DirectX SDK) starts off with callback functions that receive an `IDirect3DDevice9` reference. In our examples, this is the `pDevice` variable, which is the abstraction of the current rendering device. Methods of this interface can be used to create resources, set render states and issue draw calls. Textures can be bound to numbered samplers, vertex buffers to numbered input slots, and uniform parameters to numbered registers.

   Resources can be created in three pools: the *default pool*, the *managed pool*, and the *system memory pool*. The system memory pool may only be used for CPU–GPU communication, as it cannot be accessed by GPU shaders. Default pool resources can be imagined to reside in the GPU memory, while for managed ones, a system memory copy is maintained, and transferred to the graphics card when necessary. Thus, there can be more managed resources than available graphics memory, and when the device is lost, managed resources do not have to be recreated. However, copying to system memory and back means that resources that are computed on the GPU (render target textures) or resources that are updated frequently by the GPU cannot be managed. Resource usage flags further specify where the resource data should be stored and how it can be accessed from the CPU program. A resource in the default pool can be made dynamic, indicating that the CPU will update it regularly. It can also be made "write only" for the CPU, eliminating the need to ever copy the resource from graphics memory to system memory.

A *vertex buffer* is created by the `CreateVertexBuffer` method. This method returns a `IDirect3DVertexBuffer9` type reference that we call `pVertexBuffer`, which allows the mapping of the resource to the system memory through its `Lock` method. In accordance with the `D3DUSAGE_WRITEONLY` usage of the buffer, the flag `D3DLOCK_DISCARD` indicates that the previous contents of the buffer should not be copied back from the GPU. The system memory area can be filled to define the vertex buffer. In the following example, a vertex consists of its position, the normal vector, and the texture coordinates. Position and normal are given as the three Cartesian coordinates in modeling space. Texture coordinates use two values. A simple way to specify such a layout is using the *flexible vertex format*, or *FVF* parameter of the `CreateVertexBuffer` method, which encodes the layout in a single DWORD. In our case, it should be set to `D3DFVF_XYZ | D3DFVF_NORMAL | D3DFVF_TEX1 | D3DFVF_TEXCOORDSIZE2(0)`, indicating one set of texture coordinates of type `float2`. We set up the vertex buffer to have three vertices, i.e. to define a single triangle. Having filled the vertex buffer in the system memory, it is unlocked by the `Unlock` method. Intuitively, this is when the provided vertex data are copied to the GPU memory.

```
struct MyVertex {
    D3DXVECTOR3 position;
    D3DXVECTOR3 normal;
    D3DXVECTOR2 texcoord;
};
LPDIRECT3DVERTEXBUFFER9  pVertexBuffer;       // Pointer to the vertex buffer
pDevice->CreateVertexBuffer(
    3 * sizeof(MyVertex),                     // Byte length of buffer
    D3DUSAGE_DYNAMIC | D3DUSAGE_WRITEONLY,    // CPU access
    D3DFVF_XYZ | D3DFVF_NORMAL | D3DFVF_TEX1 | D3DFVF_TEXCOORDSIZE2(0),  // FVF
    D3DPOOL_DEFAULT,                          // Directly in graphics memory
    &pVertexBuffer,                           // Pointer to the created vertex buffer
    NULL);
MyVertex* vertices;      // Pointer to the CPU memory mapped to the buffer
pVertexBuffer->Lock(0, 0, (void**)&vertices, D3DLOCK_DISCARD); // Map to CPU memory
vertices[0].position = D3DXVECTOR3(0, 0, 0);
vertices[1].position = D3DXVECTOR3(0, 1, 0);
vertices[2].position = D3DXVECTOR3(1, 0, 0);
vertices[0].normal = vertices[1].normal = vertices[2].normal = D3DXVECTOR3(0, 0, 1);
vertices[0].texcoord = D3DXVECTOR2(0, 0);
vertices[1].texcoord = D3DXVECTOR2(0, -1);
vertices[2].texcoord = D3DXVECTOR2(1, 0);
pVertexBuffer->Unlock();                                       //Send to the GPU
```

There are two ways to define the input layout, according to which the vertex buffer should be interpreted. We have already seen the *flexible vertex format*, or *FVF*, which encodes the layout in a single DWORD. It is limited to standard layouts, elements must be ordered, it cannot be used with instancing, and it is problematic if the layout compile time is not known. However, it is convenient in the typical case when we work with a simple, fixed-layout format.

In the more general case, a *vertex format declaration* is required, which explicitly lists the elements with their associated semantics, byte widths, and offsets. For example, the following declaration means that three floats starting at offset 0 define the position, another three floats starting at offset 12 (i.e. right after the position) specify the normal vector, and the final two floats at offset 24 are the texture coordinates. The first field of an element is the buffer index, where 0 stands for the vertex buffer. The byte offset and the encoding type follow. The declaration method field could be used to specify additional processing information for the fixed-function pipeline, if we were using it. The last two fields together define the semantics.

```
D3DVERTEXELEMENT9 vertexElements[] = {
    {0,  0, D3DDECLTYPE_FLOAT3, D3DDECLMETHOD_DEFAULT, D3DDECLUSAGE_POSITION, 0},
    {0, 12, D3DDECLTYPE_FLOAT3, D3DDECLMETHOD_DEFAULT, D3DDECLUSAGE_NORMAL,   0},
    {0, 24, D3DDECLTYPE_FLOAT2, D3DDECLMETHOD_DEFAULT, D3DDECLUSAGE_TEXCOORD, 0},
    D3DDECL_END()
}; pDevice->CreateVertexDeclaration(vertexElements,
&pVertexFormatDecl);
```

The above buffer definition and format declaration can be used to render a non-indexed triangle. The rendering process is enclosed by the BeginScene and EndScene method invocations preparing the rendering device and flushing its buffered data at the end, respectively. The vertex buffer is bound to the device by the SetStreamSource method, using index 0 (index 1 will be used for the instance buffer). The input layout defined above is set using SetVertexDeclaration. The primitive type (D3DPT_TRIANGLELIST) is given as a parameter to the draw call (DrawPrimitive), not as a separate render state.

```
pDevice->BeginScene();
pDevice->SetStreamSource(0,                          // Stream index
                    pVertexBuffer,              // Vertex buffer
                    0, sizeof(MyVertex));       // Byte stride == vertex size
pDevice->SetVertexDeclaration(pVertexFormatDecl); // Set vertex layout
// or: pDevice->setFVF(D3DFVF_XYZ | D3DFVF_NORMAL);
pDevice->DrawPrimitive( D3DPT_TRIANGLELIST, // Triangle list
                    0,                       // Start at vertex 0
                    1);                      // Drawing 1 triangle
pDevice->EndScene();
```

So far we have dealt with non-indexed drawing. Now we examine how the example should be modified to use indexed drawing. The *index buffer* is created using method CreateIndexBuffer, and defined just like vertex buffers. Index buffers always contain integer indices, so no format declaration is necessary except for specifying whether indices are 16- or 32-bit integers.

```
LPDIRECT3DINDEXBUFFER9  pIndexBuffer;          // Pointer to the index buffer
pDevice->CreateIndexBuffer(sizeof(short) * 3, // Byte length of index buffer
                    D3DUSAGE_DYNAMIC | D3DUSAGE_WRITEONLY,
                    D3DFMT_INDEX16,    // 16 bit shorts
                    D3DPOOL_DEFAULT,
                    &pIndexBuffer,     // Pointer to the created index buffer
                    NULL);

unsigned short* indices;
pIndexBuffer->Lock(0, 0, (void**)&indices, D3DLOCK_DISCARD);
indices[0] = 0; indices[1] = 1; indices[2] = 2;
pIndexBuffer->Unlock();
```

The index buffer is bound using SetIndices, and an indexed triangle can be drawn with the DrawIndexedPrimitive method.

```
pDevice->BeginScene();
pDevice->SetIndices(pIndexBuffer);
pDevice->SetStreamSource(0, // Vertex buffer stream index
                    pVertexBuffer, 0,
                    sizeof(MyVertex) /* Byte stride */);
pDevice->SetVertexDeclaration(pVertexFormatDecl);
pDevice->DrawIndexedPrimitive(D3DPT_TRIANGLELIST, // Primitive type
                    0, 0, 3 /* Number of vertices */,
                    0, 1 /* Number of primitives */);
pDevice->EndScene();
```

Note that vertex and index buffers are usually filled with model data loaded from files. Parsing these files, creating, defining, binding buffers, and drawing the models are usually encapsulated in a mesh class. Direct3D provides the ID3DXMesh interface along with a set of mesh processing operations.

If we want to use instancing, a secondary vertex buffer has to be created. In the following example, we set up an instance buffer with 200 random World transformations in order to render instances of the same geometry at different locations and orientations.

```
struct MyInstanceData {
    D3DXMATRIX world;
};

LPDIRECT3DVERTEXBUFFER9 pInstanceBuffer; // Vertex buffer used as instance buffer
pDevice->CreateVertexBuffer(sizeof(MyInstanceData) * 200,  // Byte length
                        D3DUSAGE_DYNAMIC | D3DUSAGE_WRITEONLY,
                        0,                              // Not an FVF buffer
                        D3DPOOL_DEFAULT,
                        &pInstanceBuffer, NULL);

MyInstanceData* instances;
// Map instance buffer to system memory
pInstanceBuffer->Lock(0, 0, (void**)&instances, D3DLOCK_DISCARD);

#define RND ((double)rand()/RAND_MAX * 2 - 1)
for(int i=0; i<200; i++) {
    D3DXMATRIX instanceTranslation, instanceRotation;
    D3DXMatrixTranslation(&instanceTranslation, RND, RND, RND);
    D3DXMatrixRotationAxis(&instanceRotation, &D3DXVECTOR3(RND, RND, RND), RND * 3.14);
    instances[i].world = instanceRotation * instanceTranslation;
}
pInstanceBuffer->Unlock();  // Load up to GPU
```

The vertex format declaration that includes both the primary vertex and the instance buffer is shown below. The first field of every element indicates which buffer the data comes from, 0 for the vertex buffer and 1 for the instance buffer. The transformation matrix is encoded in elements associated with the TEXCOORD1, ..., TEXCOORD4 semantics.

```
D3DVERTEXELEMENT9 instancingVertexElements[] ={
    {0, 0,  D3DDECLTYPE_FLOAT3, D3DDECLMETHOD_DEFAULT, D3DDECLUSAGE_POSITION, 0},
    {0, 12, D3DDECLTYPE_FLOAT3, D3DDECLMETHOD_DEFAULT, D3DDECLUSAGE_NORMAL,   0},
    {0, 24, D3DDECLTYPE_FLOAT2, D3DDECLMETHOD_DEFAULT, D3DDECLUSAGE_TEXCOORD, 0},
    {1, 0,  D3DDECLTYPE_FLOAT4, D3DDECLMETHOD_DEFAULT, D3DDECLUSAGE_TEXCOORD, 1},
    {1, 16, D3DDECLTYPE_FLOAT4, D3DDECLMETHOD_DEFAULT, D3DDECLUSAGE_TEXCOORD, 2},
    {1, 32, D3DDECLTYPE_FLOAT4, D3DDECLMETHOD_DEFAULT, D3DDECLUSAGE_TEXCOORD, 3},
    {1, 48, D3DDECLTYPE_FLOAT4, D3DDECLMETHOD_DEFAULT, D3DDECLUSAGE_TEXCOORD, 4},
    D3DDECL_END()
};
pDevice->CreateVertexDeclaration(instancingVertexElements, &pInstancingDecl);
```

Using this buffer and declaration, we render 200 displaced instances of the same triangle in the next piece of code. It is the responsibility of the vertex shader to apply the instance transformation (passed in the TEXCOORD1, ..., TEXCOORD4 input registers) to the vertex position.

How to set up the vertex shader to do that will be discussed in Section 3.2.2. The instance buffer is bound to stream source index 1, and `SetStreamSourceFreq` is used to specify how many instances are rendered.

```
pDevice->BeginScene();
// Mark the vertex buffer as the indexed data and set instance count to 200
pDevice->SetStreamSourceFreq(0, D3DSTREAMSOURCE_INDEXEDDATA | 200);
pDevice->SetStreamSource(0, pVertexBuffer, 0, sizeof(MyVertex));
// Mark the instance buffer as instance data
pDevice->SetStreamSourceFreq(1, D3DSTREAMSOURCE_INSTANCEDATA | 1);
pDevice->SetStreamSource(1, pInstanceBuffer, 0, sizeof(MyInstanceData));
pDevice->SetVertexDeclaration(pInstancingDecl);
pDevice->SetIndices(pIndexBuffer);
pDevice->DrawIndexedPrimitive(D3DPT_TRIANGLELIST, 0, 0, 3, 0, 1);
pDevice->EndScene();
```

Note that all settings are persistent. If in a next drawing pass we do not use instancing, we have to reset defaults.

The `SetRenderState(D3DRENDERSTATETYPE State, DWORD Value)` method is used to set all *render states*, irrespectively of the controlled pipeline stage. For instance, the *depth test* can be disabled by

```
pDevice->SetRenderState(D3DRS_ZENABLE, D3DZB_FALSE);
```

The depth test is enabled again by calling the same function with D3DZB_TRUE. Identifiers such as D3DRS_ZENABLE, D3DZB_FALSE, and D3DZB_TRUE are values defined in SDK header files as preprocessor macros. Render state changes are in effect as long as the render state is not changed again.

3.2.2 Controlling the Pipeline Using Effect Files

With all these render states and uniform variables to be set, shaders to be compiled and resources bound, setting up the pipeline for a drawing pass is a lengthy process. Doing all this from the CPU program is possible, but needs the recompilation of the CPU code when anything changed. This approach also separates HLSL shader programs from the attributes governing their operation, and the operation of other stages during the same pass. The *effect framework* for HLSL offers the solution. Among other features, it allows the definition of

1. *techniques* consisting of passes, where a *pass* is a collection of applied render states and HLSL shader programs for programmable stages;

2. *texture samplers* to which a resource can be bound from the CPU program by name, and which can be used to access textures from the HLSL shader code;

3. *global variables* which may be assigned a value from the CPU program by name, and which can be accessed from HLSL shader code as uniform parameters.

In the next simple *effect file*, we define a technique with a single pass, which executes instancing of a textured triangle. HLSL shaders were detailed in Section 3.1. We use the global variable `ViewProj` and the texture sampler `ColorSampler`, both accessible as uniform parameters from any shader function. The vertex shader transforms the vertex position with the instance `World` transformation, then computes the output clipping space coordinates using the `ViewProj` matrix. The fragment shader outputs the value read from the texture bound to `ColorSampler`.

```
// Global variable accessible as a uniform parameter in all shaders
float4x4 ViewProj;          // Transformation matrix

// Texture sampler
texture2D ColorMap;          // Texture map
sampler2D ColorSampler = sampler_state {
  minfilter = linear;       // Filtering modes
  magfilter = linear;
  mipfilter = linear;
  texture = <ColorMap>;
};

// Vertex shader function in HLSL
void InstancingVS(
    in float4 Pos : POSITION,        // Input position in modeling space
    in float4 Tex : TEXCOORD0,       // Input texture coordinates
    in float4 iWorld0 : TEXCOORD1,   // Rows of the instance World transformation matrix
    in float4 iWorld1 : TEXCOORD2,
    in float4 iWorld2 : TEXCOORD3,
    in float4 iWorld3 : TEXCOORD4,
    out float4 hPos : POSITION,      // Output position in clipping space
    out float2 oTex : TEXCOORD0      // Output texture coordinates
) {
    // Assemble instance World matrix
    float4x4 iWorld = float4x4(iWorld0, iWorld1, iWorld2, iWorld3);
    hPos = mul(mul(Pos, iWorld), ViewProj); // Transform vertex to clipping space
    oTex = Tex;
}

// Pixel shader function in HLSL
```

```
float4 TexturingPS( in float2 Tex : TEXCOORD0 ) : COLOR0
{
    return tex2D(ColorSampler, Tex);
}


// Technique definition
technique ExampleTechnique {
    pass ExamplePass {
        // Render states
        ZEnable = TRUE;             // Enable depth buffering
        FillMode = SOLID;           // Fill triangles
        // Shaders compiled for Shader Model 3
        VertexShader = compile vs_3_0 InstancingVS();
        PixelShader  = compile ps_3_0 TexturingPS();
    }
}
```

The keywords for render states and their possible values are as numerous as render states themselves, and they are listed in the SDK reference. The above pass may be initiated from the CPU program using the `ID3DXEffect` interface. The effect defined in file "PipeControl.fx" is compiled with:

```
ID3DXEffect* pEffect; // D3DX effect interface
LPD3DXBUFFER pErrors;  // Buffer wrapping compilation error string
DXUTFindDXSDKMediaFileCch(str, MAX_PATH, L"PipeControl.fx" );
D3DXCreateEffectFromFile(pDevice, str, NULL, NULL, 0, NULL, &pEffect, &pErrors);
```

How the scene with these effect settings can be rendered is shown below, assuming the vertex, index, and instance buffers are already set up as detailed previously in Section 3.2.1. Various methods of `ID3DXEffect` allow the setting of global variables, binding of textures, and choosing of the appropriate technique and pass. We assume the CPU program variables `ViewProj` of type `D3DXMATRIX` and `pTexture` of type `LPDIRECT3DTEXTURE9` have been defined and set appropriately.

```
pEffect->SetMatrix("ViewProj", &ViewProj);    // Set View-Projection matrix
pEffect->SetTexture("ColorMap", pTexture);    // Bind texture to sampler
pEffect->CommitChanges();                      // Apply changes
pDevice->BeginScene();
pEffect->SetTechnique("ExampleTechnique");    // Choose technique
UINT a;
pEffect->Begin(&a, 0);      // Start technique querying the number of passes
pEffect->BeginPass(0);      // Initialize 0th pass
pDevice->DrawIndexedPrimitive( D3DPT_TRIANGLELIST, 0, 0, 3, 0, 1);
```

```
pEffect->EndPass();
pEffect->End();
pDevice->EndScene();
```

Thus, the effect framework removes the burden of managing render states and associating shader input resources and uniform parameters with hardware units and registers. Creating shader input resources and binding them to effect handles remain the CPU tasks, just like setting the input vertex buffer, instance buffer, index buffer, and output render targets and buffers.

3.2.2.1 Controlling the Pipeline From Direct3D 10

In Direct3D 10, render state is organized into *state objects*, basically corresponding to pipeline stages. Immutable instances of these state structures can be defined matching the scenarios emerging in the application. A state change means switching from one predefined instance to another, without requiring the manual change of state elements one by one. Creating state objects and setting render states are done using methods of the `ID3DDevice` interface.

The input assembler stage does not have state objects, but *input layouts* come close. They replace vertex declarations of Direct3D 9, listing elements in a very similar manner. However, an input layout also contains the information about the mapping of buffer elements to shader inputs: something that varies for shaders using different non-uniform input parameters. Therefore, when creating an input layout, we must also provide the shader input signature returned by the shader compiler. The primitive type for the input assembler can be specified using `IASetPrimitiveTopology`, and it is not a parameter of a draw call any more. Vertex and instance buffers are bound using `IASetVertexBuffers`. For the index buffer we can use `IASetIndexBuffer`.

Rasterizer state objects are created by method `CreateRasterizerState`, and set by `RSSetState`. The output merger stage state is divided into two: depth stencil state and blend state.

Draw calls are leaner. They require neither the primitive type, nor the vertex buffer stride in bytes as parameters. There is an indexed and non-indexed method, and a version with instancing for both. No manual tinkering with input stream frequencies is required any more. Furthermore, Shader Model 4.0 shaders may receive system-generated values as input parameters, including the vertex index and the instance index. Thus, it is not necessary to store this information in the vertex buffer.

Effect files are still used. However, the syntax has been changed to remove the differences between API render state and shader definitions. Render state objects and variables referencing shaders can be defined, and they are set in the pass definition with functions identical to the API device methods. The following is an example fragment from an effect file, defining render state and shader objects, and using them to set up a technique with a pass. We assume that the top-level functions called `InstancingVS` and `TexturingPS` have been defined. These should be similar to the Shader Model 3 shaders used in the previous section, but with Shader Model 4 output semantics: `SV_TARGET` replaces `COLOR` and `SV_POSITION` replaces `POSITION`.

```
// An immutable state block object
BlendState SrcAlphaBlendingAdd {
// State names are identical to API field names
    BlendEnable[0] = TRUE;
    SrcBlend = SRC_ALPHA;
    DestBlend = ONE;
    BlendOp = ADD;
    SrcBlendAlpha = ZERO;
    DestBlendAlpha = ZERO;
    BlendOpAlpha = ADD;
    RenderTargetWriteMask[0] = 0x0F;
};


// Compiled shaders are also objects
PixelShader psObject = CompileShader(ps_4_0, TexturingPS());


technique10 ExampleTechnique {
    pass ExamplePass {
        // Render states and shaders are set with API functions
        SetBlendState(SrcAlphaBlendingAdd, float4(0 , 0, 0, 0), 0xFFFFFFFF);
        SetVertexShader( CompileShader(vs_4_0, InstancingVS()) );
        SetGeometryShader( NULL );
        SetPixelShader( psObject );
    }
}
```

3.3   SHADERS BEYOND THE STANDARD PIPELINE OPERATION

Programmable vertex and fragment shaders offer a higher level of flexibility on how the data from the vertex buffer are processed, and how shading is performed. However, the basic pipeline model remains the same: a vertex is processed, the results are linearly interpolated, and they are used to find the color of a fragment. The flexibility of the programmable stages will allow us to change the illumination model: implement per-pixel lighting (Figure 2.8), consider more and different light sources, combine several textures, and much more.

3.3.1 Are GPUs Good for Global Illumination?

What programmable vertex and fragment shaders alone cannot do is non-local illumination. All the data passed to shaders are still only describing local geometry and materials, or global constants, but nothing about the other pieces of geometry. When a point is shaded with a global illumination algorithm, its radiance will be the function of all other points in the scene. From a programming point of view, this means that we need to access the complete scene description when shading a point. While this is granted in CPU based ray–tracing systems [151, 150], the stream processing architecture of current GPUs fundamentally contradicts this requirement. When a point is shaded on the GPU, we have just its limited amount of local properties stored in registers, and may access texture data. Thus, the required global properties of the scene must be stored in *textures*.

If the textures must be static, or they must be computed on the CPU, then the lion's share of illumination does not make use of the processing power of the parallel hardware, and the graphics card merely presents CPU results. Textures themselves have to be computed on the GPU. The *render-to-texture* feature allows this: anything that can be rendered to the screen, may be stored in a texture. Such texture render targets may also require depth and stencil buffers. Along with programmability, various kinds of data may be computed to textures. They may also be stored in floating point format in the texture memory, unlike the color buffer which usually stores data of 8 bit precision.

To use textures generated by the GPU, the rendering process must be decomposed to passes, also called *rendering runs*, where one run may render into a texture and another may use the generated textures (in effect framework terms, rendering runs are passes of different techniques). Since the reflected radiance also depends on geometric properties, these textures usually contain not only conventional color data, but they also encode geometry and prepared, reusable illumination information as well.

Programmability and render-to-texture together make it possible to create some kind of processed representation of geometry and illumination as textures. These data can be accessed when rendering and shading other parts of the scene. This is the key to addressing the self-dependency of the global illumination rendering problem. In all GPUGI algorithms, we use multiple passes to different render targets to capture some aspects of the scene like the surrounding environment, the shadowing or the refracting geometry, illumination due to light samples, etc. These passes belong to the *illumination information generation* part of rendering (Figure 3.1). In a final pass, also called *final gathering*, scene objects are rendered to the frame buffer making use of previously computed information to achieve non-local illumination effects like shadows, reflections, caustics, or indirect illumination.

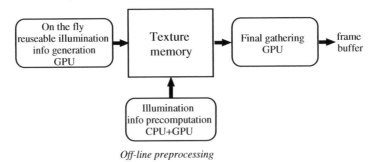

Off-line preprocessing

FIGURE 3.1: Structure of real-time global illumination shaders.

Passes of the illumination information generation part are responsible for preparing the reusable illumination information and storing it in the texture memory, from which the final gathering part produces the image for the particular camera. To produce continuous animation, the final gathering part should run at high frame rates. Since the scene may change, the illumination information generation should also be repeated. However, if the illumination information is cleverly defined, then its elements can be reused for the shading of surface points over several frames. Thus, the illumination information data structure is compact and might be regenerated at significantly lower frequency than the final gathering frame rate.

3.3.2   Basic Techniques for GPU Global Illumination

Implementing the above scheme of using textures for global illumination straightforwardly, we may render the surroundings of a particular reflective scene entity to a cube map texture. Then, when drawing the entity, the fragment shader may retrieve colors from this texture based on the reflected eye vector (computed from the local position and normal, plus the global eye position). The structure is shown in Figure 3.2. This is the image synthesis technique known as *environment mapping* (Figure 3.3), and we describe it in detail in Section 4.2.

It is also possible to write a vertex shader that exchanges the texture and position co-ordinates. When drawing a model mesh, the result is that the triangles are rendered to their positions in texture space. Anything computed for the texels may later be mapped on the

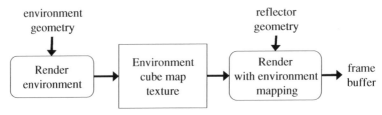

FIGURE 3.2: Dataflow structure of environment mapping.

FIGURE 3.3: Environment mapping using a metal shader.

mesh by conventional texture mapping. Figure 3.4 depicts the process. This method assumes that the mapping is unique, and such a render texture resource is usually called a *UV atlas* (Figure 3.5). In this book, whenever we use the term *atlas* or *texture atlas*, we refer to such a unique mapping. Note that other sources also use *texture atlas* in any case where logically separated data are stored in a single texture, most notably when tiling small images into one large texture resource.

The simplest vertex shader rendering to a UV atlas is given below. In practical applications, the vertex shader also performs other computations and passes more data to the fragment shader, depending on the specific rendering task. In Sections 6.7 and 8.4.1, we will describe such techniques in detail.

```
void ToAtlasVS(
    in float2 AtlasTex : TEXCOORD0,
    out float4 hPos : POSITION
) {
```

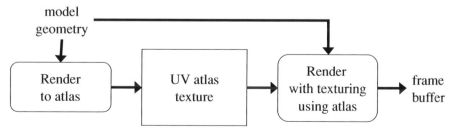

FIGURE 3.4: Dataflow structure of texture atlas generation and usage.

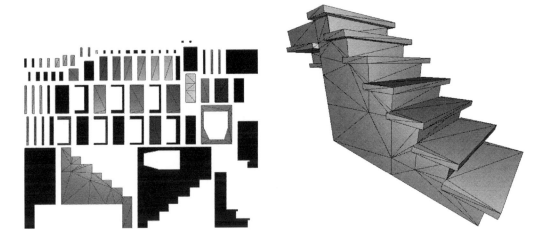

FIGURE 3.5: A staircase model rendered to UV atlas space (left) and with the conventional 3D transformation pipeline (right).

```
    // Texture space to screen space
    float2 xyPos = AtlasTex * float2(2, -2) - float2(1, 1);
    hPos = float4(xyPos, 0, 1);
}
```

When additional textures are used to achieve global illumination effects, it becomes a necessity to be able to access multiple textures when computing a fragment color. This is called *multi-texturing*. Even with Shader Model 3.0, fragment shaders are able to access as many as 16 textures, any one of them multiple times. They all can be addressed and filtered using different modes or sets of texture coordinates, and their results can be combined freely in the fragment program. This allows the simultaneous use of different rendering techniques such as environment mapping, bump mapping, light mapping, shadow mapping, etc.

It is also a likely scenario that we need to compute multiple values for a rendering setup. This is accomplished using *multiple render targets*: a fragment shader may output several values (pixel colors) that will be written to corresponding pixels of respective render targets. With this feature, computing data that would not fit in a single texture becomes feasible. For instance, *deferred shading* renders all visible geometry and material properties into screen-sized textures, and then uses these textures to render the shaded scene without actually rendering the geometry. A fragment shader writing to multiple render targets simply defines more out parameters with semantics ranging from COLOR0 to COLOR3.

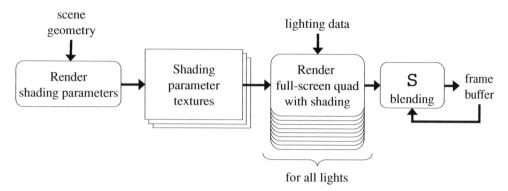

FIGURE 3.6: Dataflow structure of deferred shading with a high number of light sources.

The following fragment shader example could be used in a naive, non-texturing deferred shading implementation. The position and normal textures to be used in later passes must be assigned as render targets.

```
void MultipleRenderTargetPS(
    in float4 wPos : TEXCOORD0,    // World position
    in float3 Norm : TEXCOORD1,    // World normal
    out float4 oPos : COLOR0,      // First target:  position
    out float4 oNormal : COLOR1    // Second target: normal
) {
    oPos = hPos;
    oNormal = float4(Norm, 0);
}
```

Figure 3.6 shows the dataflow structure of deferred shading. The complex scene geometry is only drawn once, filling textures with the local properties necessary for shading computations. Then, the contribution of any number of light sources (or arbitrary effects) can be rendered by drawing full-screen quads consecutively, without the need to process the actual geometry.

As we shall discuss in this book, the features and techniques detailed above give us enough freedom to implement global illumination algorithms and even ray–tracing approaches.

CHAPTER 4

Simple Improvements of the Local Illumination Model

Local illumination models simplify the rendering problem to shading a surface fragment according to a given point or directional light source. This is based on several false assumptions resulting in less realistic images:

The light source is always visible. In reality, incoming lighting depends on the materials and geometry found between the light source and the shaded point. Most prominently, solid objects may occlude the light. Neglecting this effect, we will render images without *shadows*. In order to eliminate this shortcoming, we may capture occluding geometry in texture maps and test the light source for visibility when shading. This technique is called *shadow mapping*.

The light illuminates from a single direction. In reality, light emitters occupy some volume. The point light source model is suitable for small artificial lights like a bulb. The directional light source model is good for large distance sources like the Sun. However, in most environments we encounter extended light sources that do not belong to these models. The sky itself is a huge light source. In *image-based lighting*, we place virtual objects in a computed or captured environment and compute the illumination of this environment on virtual objects. Volumetric or area lights generate more elaborate shadows, as they might be only partly visible from a given point. These shadows are often called *soft shadows*, as they do not feature a sharp boundary between shadowed and lighted surfaces. Generally, multi-sampling of extended light sources is required to render accurate shadows. However, with some simplifying assumptions for the light source, faster approximate methods may be obtained, generating perceptionally plausible soft shadows.

No indirect lighting. In reality, all illuminated objects reflect light, lighting other objects. While this indirect lighting effect constitutes a huge fraction of light we perceive, it tends to be at low frequency and is less obvious, as most surfaces scatter the light diffusely. However, for highly specular, metallic, mirror-like or refractive materials, this does not apply. Indirect illumination may exhibit elaborate high-frequency patterns called caustics, and of course the color we see on a mirror's surface depends on the surrounding geometry.

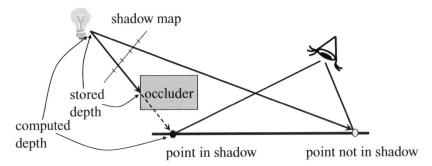

FIGURE 4.1: Shadow map algorithm. First the scene is rendered from the point of view of the light source and depth values are stored in the pixels of the generated image, called shadow map or depth map. Then the image is taken from the real camera, while points are also transformed to the space of the shadow map. The depth of the point from the light source is compared to the depth value stored in the shadow map. If the stored value is smaller, then the given point is not visible from the light source, and therefore it is in shadow.

These issues require more sophisticated methods, based on the approach of *environment mapping*, capturing incoming environment radiance in textures.

4.1   SHADOW MAPPING

Shadows are important not only to make the image realistic, but also to allow humans to perceive depth and distances. Shadows occur when an object called *shadow caster* occludes the light source from another object, called *shadow receiver*, thus prevents the light source from illuminating the shadow receiver. In real-time applications shadow casters and receivers are often distinguished, which excludes *self-shadowing* effects. However, in real life all objects may act as both shadow receiver and shadow caster.

Point and directional light sources generate *hard shadows* having well-defined boundaries of illumination discontinuities. Real-time shadow algorithms can be roughly categorized as image space *shadow map* [156] or object space *shadow volume* [28] techniques. Shadow volumes require geometric processing, as they construct invisible faces to find out which points are in the shadow. Exact geometric representation allows exact shadow boundaries, but the computation time grows with the geometric complexity, and partially transparent objects, such as billboards become problematic.

The shadow map algorithm, on the other hand, works with a depth image (also called *depth map* or *shadow map*), which is a sampled version of the shadow caster geometry. Since shadow map methods use only a captured image of the scene as seen from the light, they are independent of the geometric complexity, and can conveniently handle *transparent* surfaces as well. However, their major drawback is that the shadowing information is in a discretized form, as a collection of shadow map pixels called *lexels*, thus sampling or aliasing artifacts are likely to occur.

For the sake of simplicity, we assume that the light sources are either *directional* or *spot* lights having a main illumination direction. *Omnidirectional* lights may be considered as six spot lights radiating towards the six sides of a cube placed around the omnidirectional light source.

Shadow map algorithms use several coordinate systems:

World space: This is the arbitrary global frame of reference for specifying positions and orientations of virtual world objects, light sources, and cameras. Object models are specified in their own reference frame using modeling coordinates. For an actual instance of a model, there is a transformation that moves object points from modeling space to world space. This is typically a homogeneous linear transformation, called the modeling transformation or `World`.

Eye's camera space: In this space, the eye position of the camera is at the origin, the viewing direction is the z-axis in Direct3D and the $-z$ direction in OpenGL, and the vertical direction of the camera is the y-axis. Distances and angles are not distorted, lighting computations can be carried out identically to the world space. The transformation from modeling space to this space is `WorldView`.

Light's camera space: In this space the light is at the origin, and the main light direction is the z-axis. This space is similar to the eye's camera space having replaced the roles of the light and the eye. The transformation from modeling space to this space is `DepthWorldView`, which must be set according to the light's position.

Eye's clipping space: Here the eye is at an ideal point $[0, 0, 1, 0]$, thus the viewing rays get parallel. The potentially visible part of the space is an axis aligned box of corners $(-1, -1, 0)$ and $(1, 1, 1)$ in the Cartesian coordinates assuming Direct3D. The transformation to this space is not an *affine transformation*[1], thus the fourth homogeneous coordinate of the transformed point is usually not equal to 1. The transformation from modeling space to this space is `WorldViewProj`.

Light's clipping space: Here the light is at an ideal point $[0, 0, 1, 0]$, thus the lighting rays get parallel. The potentially illuminated part of the space is an axis aligned box of corners $(-1, -1, 0)$ and $(1, 1, 1)$ in the Cartesian coordinates. The transformation to this space is not an affine transformation, thus the fourth homogeneous coordinate is usually not equal to 1. The transformation from modeling space to this space is `DepthWorldViewProj`. This matrix should be set according to light characteristics. For a directional light, the light rays

---

[1]A transformation is said to be affine if it maps parallel lines to parallel lines.

are already parallel, so we only need to translate the origin to the light source position, rotate the scene to make the main light direction parallel with the axis *z*, and apply a scaling to map the interesting, potentially illuminated objects into the unit box. For point lights, a perspective projection matrix is needed, identical to that of perspective cameras. The field of view should be large enough to accommodate for the spread of the spotlight. Omnidirectional lights need to be substituted by six 90° *field of view* lights.

Shadow mapping has multiple passes, a shadow map generation pass for every light source when the camera is placed at the light, and an image generation pass when the camera is at the eye position.

4.1.1   Shadow Map Generation

Shadow map generation is a regular rendering pass where the z-buffer should be enabled. The actual output is the z-buffer texture with the depth values. Although a color target buffer is generally required, color writes should be disabled in the effect file to increase performance. Typically the following render states are set:

```
ZEnable          = True;   // Enable depth buffering
Lighting         = False;  // Disable per-vertex illumination computation
ColorWriteEnable = False;  // No need to render to the color buffer, we only need z
```

Transformation `DepthWorldViewProj` is set to transform points to the world space, then to the light's camera space, and finally to the light's clipping space. This matrix is obtained by first computing the transformation from world space to light's camera space from light position `lightPos`, light direction (`lightLookAt` - `lightPos`), and the vertical direction associated with the light's camera, `lightUp`. The perspective transformation from the light's camera space to its clipping space is produced from the light's field of view (`lightFov`) and the front (`lightFront`) and back (`lightBack`) clipping plane distances:

```
D3DXMATRIX DepthWorldView, LightView, DepthWorldLightProj;
D3DXMatrixLookAtLH(&LightView, &lightPos, &lightLookAt, &lightUp);
DepthWorldView = World * LightView; // from modeling to light's camera space
D3DXMatrixPerspectiveFovLH(&LightProj,D3DXToRadian(lightFov),1,lightFront,lightBack);
DepthWorldViewProj = DepthWorldView * LightProj;
```

The shader executes a regular rendering phase. Note that the pixel shader color is meaningless since it is ignored by the hardware anyway. The effect file containing the vertex and fragment shaders are as follows:

```
float4x4 DepthWorldViewProj; // From model space to light's clipping space

// Vertex shader of depth map creation
void DepthVS( in  Pos  : POSITION,  // Input position in modeling space
              out hPos : POSITION   // Output position in light's clipping space
            )
{
    hPos = mul(Pos, DepthWorldViewProj);
}

float4 DepthPS( ) : COLOR0 { return 0; }  // Fragment shader
```

4.1.2 Rendering With the Shadow Map

In the second phase, the scene is rendered from the eye's camera. Simultaneously, each visible point is transformed to the light's clipping space, then to the texture space where depth values are stored, and its depth value is compared to the stored depth value. The texture space transformation is responsible for mapping spatial coordinate range $[-1, 1]$ to $[0, 1]$ texture range, inverting the y coordinate, since the spatial y coordinates increase from bottom to top, while the texture coordinates increase from top to bottom. Let us denote the texture address by u, v and the associated depth by d. When the texture address is computed, we should shift the u, v texture address by half a texel since texel centers are half a texel away from their top left corner. The depth value is used for comparison to decide whether the same point is stored in the shadow map. Ideally, if a point is not occluded from the light source, then its stored depth value is identical to the depth computed during the camera pass. However, due to numerical inaccuracies and to the final resolution of the depth map, it might happen that the stored depth is a little smaller. In this case, the precise comparison would decide that the surface occludes itself. To avoid incorrect *self-shadowing*, a small bias is added to the depth value before comparison.

Note that in clipping space the point is given by homogeneous coordinates $[x, y, z, w]$, from which the homogeneous division generates the Cartesian triplet $(x/w, y/w, z/w)$. Thus, the transformation step generating texture coordinates and the biased depth is

$$u = (x/w + 1)/2 + h,$$
$$v = (-y/w + 1)/2 + h,$$
$$d = z/w - b,$$

where h is half a texel and b is the *depth bias*. If the shadow map resolution is SHADOWMAP_SIZE, then the half texel has size $0.5/$SHADOWMAP_SIZE. Multiplying these equations by w, we can express the texture address and the biased depth in homogeneous coordinates:

$$[uw, vw, dw, w] = [x/2 + w(b + 1/2), -y/2 + w(b + 1/2), z - wb, w].$$

The texture space transformation can also be implemented as a matrix multiplication. To produce `DepthWorldViewProjTex`, the following `TexScaleBias` matrix must be appended to the transformation to light's clipping space:

$$\mathbf{T}_{Tex} = \begin{bmatrix} 1/2 & 0 & 0 & 0 \\ 0 & -1/2 & 0 & 0 \\ 0 & 0 & 1 & 0 \\ H & H & -b & 1 \end{bmatrix}, \tag{4.1}$$

where $H = b + 1/2$. The vertex shader executes this transformation in addition to normal tasks such as copying the input color (or texture address) and transforming the vertex to homogeneous clipping space:

```
// Transformation from modeling space to eye's clipping space
float4x4 WorldViewProj;

// Transformation from modeling space to texture space of the shadow map
float4x4 DepthWorldViewProjTex;

void ShadowVS(
    in  float4 Pos      : POSITION,   // Input position in modeling space
    in  float4 Color    : COLOR0,     // Input color
    out float4 hPos     : POSITION,   // Output position in eye's clipping space
    out float4 depthPos : TEXCOORD0,  // Output position in texture space
    out float4 oColor   : COLOR0)     // Output color
{
    oColor = Color; // Copy color
    // Transform model space vertex position to eye's clipping space:
    hPos = mul(Pos, WorldViewProj);
    // Transform model space vertex position to the texture space of the shadow map
    depthPos = mul(Pos, DepthWorldViewProjTex);
}
```

In the fragment shader, we check whether or not the stored depth value is smaller than the given point's depth. If it is smaller, then there is another object that is closer to the light source, thus, our point is in shadow. These operations, including the texture lookup using homogeneous texture coordinates and the comparison, are so important that GPU vendors implemented them directly in hardware. This hardware can be activated by specifying the texture storing depth values as *depth stencil texture* in Direct3D and fetching the texture value by `tex2Dproj` in HLSL. The following fragment shader uses this method and the color is multiplied by the returned visibility factor (`vis`) to make shadowed fragments black:

```
texture  ShadowMap; // Depth map in texture

sampler ShadowMapSampler = sampler_state {
    Texture = <ShadowMap>;
    MinFilter = POINT;
    MagFilter = POINT;
};

float4 ShadowPS(
    float4 depthPos : TEXCOORD0, // Texture coordinate of the shadow map
    float4 Color    : COLOR0     // Input color assuming no shadowing
    ) : COLOR
{
    // POINT sampling of a depth stencil texture returns 0 or 1 as a comparison result
    float vis = tex2Dproj(shadowMapSampler, depthPos).r;
    return vis * Color;          // Change input color to black in shadow
}
```

The code for the shadow map query is virtually identical to the code for projective textures. Projective texturing function `tex2Dproj(sampler, p)` divides `p.x, p.y, p.z` by `p.w` and looks up the texel addressed by (`p.x/p.w, p.y/p.w`). Shadow maps are like other projective textures, except that in projective texture lookups instead of returning a texture color, `tex2Dproj` returns the boolean result of the comparison of `p.z/p.w` and the value stored in the texel. To force the hardware to do this, the associated texture unit should be configured by the application for depth compare texturing; otherwise, no depth comparison is actually performed. In Direct3D, this is done by creating the texture resource with the usage flag `D3DUSAGE_DEPTHSTENCIL`. Note that `tex2D` will not work on this texture, only `tex2Dproj` will.

Classical shadow mapping requires just a single texture lookup in the pixel shader. This naive implementation has many well-known problems, caused by storing only sampled information in the depth map. These problems include *shadow acnes* caused by numerical inaccuracies and *aliasing* due to limited shadow map resolution.

Shadow acnes can be reduced by applying depth bias, but it is often a delicate issue to choose an appropriate bias for a given scene. Values exceeding the dimensions of geometric details will cause *light leaks*, non-shadowed surfaces closely behind shadow casters (left of Figure 4.6). With a bias not large enough, *z-fighting* will cause interference-like shadow stripes on lighted surfaces. Both artifacts are extremely disturbing and unrealistic. The issue might be more severe when the depth map is rendered with a large field of view angle, as depth distortion can be extreme. A convenient solution is the *second depth value* technique. Assuming that all shadow casters are non-intersecting, opaque manifold objects, the back-faces cast the same shadow as the object itself. By reversing the *back-face culling* mechanism, we can render only the back-faces, so depth values in the depth map do not coincide with any front face depth.

4.1.3   Shadow Map Anti-aliasing

All shadow map techniques suffer from aliasing artifacts due to undersampling. If lexels are large in world space then their boundaries can be recognized (Figure 4.2). This problem is made even worse by the fact that the surface area that is visible in a single lexel can be visible in multiple pixels from the camera, which makes these artifacts quite disturbing. The causes of this magnification are that the surface is far from the light source but is close to the camera (*perspective aliasing*), and that the surface is seen at grazing angles from the light, but at perpendicular direction from the camera (*projection aliasing*).

The size of lexels can be reduced by increasing the shadow map resolution even beyond the maximal texture size [43] or by *focusing the shadow map* to that part of the scene which is visible from the camera [17] (Figure 4.4). To reduce perspective aliasing, the scene should be distorted by a homogeneous transformation before the shadow map is computed. This transformation magnifies those parts of the scene that are close to the camera [130, 158].

FIGURE 4.2: Hardware shadow mapping with 512×512 shadow map resolution.

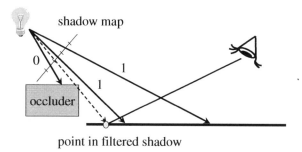

point in filtered shadow

FIGURE 4.3: Percentage closer filtering reads the shadow map at the four corners of the lexel containing the test point, and the results of the four comparisons (either 1 or 0) are bi-linearly filtered.

Aliasing can also be reduced by low-pass filtering. GPUs have several filtering methods for color textures, including the bi-linear interpolation of four texels, *mipmapping*, and anisotropic filtering (more sophisticated filtering schemes are discussed in Section 11.1). However, these filtering schemes cannot be directly applied to shadow maps, since they would interpolate the depth values of neighboring texels. If a part of the area belonging to a single screen pixel is in shadow while the rest is not in shadow, we would need the ratio of these areas to filter the pixel, and not the result of a binary decision that compares the current depth to the average depth corresponding to this pixel. If we assume that points belonging to a pixel are selected randomly with uniform distribution, then filtering should approximate the probability that the point is not in shadow. If the depth of the current point is z and the random variable of stored depths in the direction of the points belonging to the current pixel is Z, then we need probability $P(Z > z)$.

Percentage closer filtering (PCF) proposed by Reeves et al. [112] averages the boolean values representing shadow decisions. A depth comparison may result in "not visible" (say 0) or "visible" (say 1). In the PCF shadow map, comparisons are made in several close lexels and the 0, 1 values are averaged together. The reason of the name of this method is that it approximates the percentage of the surface projected onto the lexel that is not in shadow (Figure 4.3).

NVIDIA GPUs have built-in percentage-closer filtering. If the depth texture is associated with linear filtering, then the hardware automatically looks up for depth values, executes four depth compares, and applies *bi-linear filtering* to the resulting binary values (Figure 4.4). Thus, the simplest solution is identical to that of the previous subsection, but changing the filters to linear:

```
sampler ShadowMapSampler = sampler_state {
    Texture = <ShadowMap>;
    MinFilter = LINEAR;     // Changing from POINT to LINEAR turns PCF on
    MagFilter = LINEAR;     // Changing from POINT to LINEAR turns PCF on
```

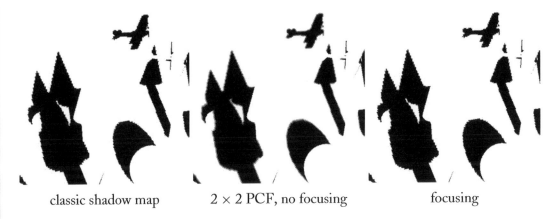

classic shadow map 2×2 PCF, no focusing   focusing

FIGURE 4.4: Comparison of point sampling (left) and bi-linear filtering, i.e. 2×2 percentage closer filtering (middle), and the focusing (right) of shadow maps.

```
};

float4 ShadowPS(
    float4 depthPos : TEXCOORD0, // Texture coordinate of the shadow map
    float4 Color    : COLOR0     // Input color assuming no shadowing
    ) : COLOR
{   // LINEAR sampling interpolates these 0/1 values, not the depth
    float vis = tex2Dproj(shadowMapSampler, depthPos).r;
    return vis * Color;
}
```

Variance shadow maps [34] exploit the filtering hardware designed for color textures. Since simple shadow map filtering would not result in useful data, this method filters independently two values, the real depth and its square. Since filtering computes a weighted average, which can be interpreted as an expected value taking the filter kernel as the probability density of a random variable, the two filtered values are mean $E[Z]$ and second moment $E[Z^2]$, respectively, where Z is the random depth. The required probability $P(Z > z)$ for a particular pixel point in depth z can be approximated using the *one tailed version of the Chebychev's inequality*[2] if $z > E[Z]$:

$$P(Z \geq z) \leq \frac{\sigma^2}{\sigma^2 + (z - E[Z])^2} \quad \text{where } \sigma^2 = E[Z^2] - E^2[Z] \text{ is the variance.}$$

---

[2]Note that the one-tailed version is sharper than the classic Chebychev's inequality. Its proof is based on the *Markov inequality* that states $P(\xi \geq a) \leq E[\xi]/a$ for a non-negative random variable ξ and positive constant a, substituting $\xi = (Z - E[Z] + \sigma^2/(z - E[Z]))^2$ and $a = (z - E[Z] + \sigma^2/(z - E[Z]))^2$.

FIGURE 4.5: Comparison of classic shadow map (left) and the variance shadow map algorithm (right). We used a 5×5 texel filter to blur the depth and the square depth values.

If $z \le E[Z]$, then the Chebychev's inequality cannot be used, but the point is assumed to be not in shadow. Variance shadow mapping reads values $E[Z]$, $E[Z^2]$ from the filtered two-channel depth map, and replaces the comparison by the upper bound of the probability.

```
float4 VarianceShadowPS(
    float4 depthPos : TEXCOORD0, // Texture coordinate of the shadow map
    float4 Color    : COLOR0     // Input color assuming no shadowing
    ) : COLOR
{
    float z = depthPos.z / depthPos.w;
    float mean = tex2D(shadowMapSampler, depthPos).r;
    float secondmoment = tex2D(shadowMapSampler, depthPos).g;
    float variance = secondmoment - mean * mean;
    float prob = variance / (variance + (z - mean) * (z - mean));
    return max(z <= mean, prob) * Color; // if z < E[z], then no shadowing
}
```

4.2 IMAGE-BASED LIGHTING

In many computer graphics applications, it is desirable to augment the virtual objects with images representing a real environment (sky, city, etc.). The radiance values representing environment illumination may differ by orders of magnitude, thus they cannot be mapped to eight bits. Instead, the red, green, and blue primaries of the pixels in these images should be stored as floating point values to cope with the high range. Floating point images are called *high dynamic range images* (Section 11.3).

FIGURE 4.6: Comparison of classic shadow map (left) and *light space perspective shadow map* [158] combined with the variance shadow map algorithm. Note that in large scenes classic shadow map makes unacceptable aliasing errors (the lexels are clearly visible) and light leak errors (the shadow leaves its caster, the column).

In order to provide the illusion that the virtual objects are parts of the real scene, the illumination of the environment should be taken into account when the virtual objects are rendered. Since the images representing the environment lack geometric information, we usually assume that the illumination stored in images comes from very (infinitely) far surfaces. This means that the environment illumination is similar to directional lights, it has only directional characteristics, but its intensity is independent of the location of the illuminated point.

A natural way of storing the direction-dependent map is an angular mapped floating point texture. Direction ω' is expressed by spherical angles θ', ϕ' where $\phi' \in [0, 2\pi]$ and $\theta' \in [0, \pi/2]$ in the case of hemispherical lighting and $\theta' \in [0, \pi]$ in the case of spherical lighting. Then the texture coordinates $[u, v]$ are scaled from the unit interval to these ranges. For example, in the case of spherical lighting, a direction is parameterized as

$$\omega' = (\cos 2\pi u \sin \pi v, \ \sin 2\pi u \sin \pi v, \ \cos \pi v),$$

where $u, v \in [0, 1]$.

Another, more GPU friendly possibility is to parameterize the directional space as sides of a cube centered at the origin and having edge size 2. A point x, y, z on the cube corresponds to direction

$$\omega' = \frac{(x, y, z)}{\sqrt{x^2 + y^2 + z^2}}.$$

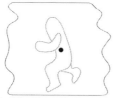 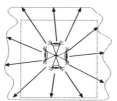 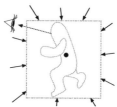

1. Finding the center
of the object

2. Taking images
from the center

3. Illumination
form the images

FIGURE 4.7: Steps of environment mapping.

One of the three coordinates is either 1 or -1. For example, the directions corresponding to the $z = 1$ face of the cube are

$$\omega' = \frac{(x, y, 1)}{\sqrt{x^2 + y^2 + 1}}, \quad x, y \in [-1, 1].$$

Current GPUs have built-in support to compute this formula and to obtain the stored value from one of the six textures of the six *cube map* faces (texCUBE in HLSL).

The very same approach can also be used for purely virtual scenes if they can be decomposed to smaller dynamic objects and to their large environment that is far from the dynamic objects (such distinction is typical in games and virtual reality systems). In this case, the illumination reflected off the environment is computed from a *reference point* placed in the vicinity of the dynamic objects, and is stored in images. Then the environment geometry is replaced by these images when dynamic objects are rendered (Figure 4.7). These images form the *environment map*, and the process is called *environment mapping* [16].

Images forming the environment map correspond to the sides of a *cube map*, since it is rather convenient to render the environment from the reference point setting the faces of a cube as windows. So the creation of the environment map requires the rendering of the environment geometry six times.

In order to compute the image of a virtual object under infinitely far environment illumination, we should evaluate the *reflected radiance L^r* due to the environment illumination at every visible point $\mathbf{x}$ at view direction ω (Figure 4.8):

$$L^r(\mathbf{x}, \omega) = \int_{\Omega'} L^{\text{env}}(\omega') f_r(\omega', \mathbf{x}, \omega) \cos^+ \theta'_{\mathbf{x}} v(\mathbf{x}, \omega') \, d\omega', \qquad (4.2)$$

where $L^{\text{env}}(\omega')$ is the radiance of the environment map at direction ω', and $v(\mathbf{x}, \omega')$ is the *visibility factor* checking whether no virtual object is seen from $\mathbf{x}$ at direction ω' (that is, the environment map can illuminate this point from the given direction). Note that the assumption that illumination arrives from very distant sources allowed the elimination of the positional

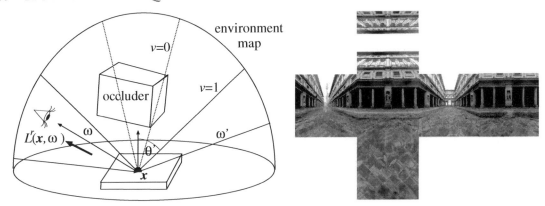

FIGURE 4.8: The concept of environment mapping and the environment map of Florence [32] stored in a cube map.

dependence from the incoming radiance and its replacement by direction-dependent environment radiance $L^{\text{env}}(\omega')$.

The illumination of the environment map on the virtual objects can be obtained by tracing rays from the points of the virtual object in the directions of the environment map, and checking whether or not occlusions occur [30, 80]. The computation of the visibility factor, that is the shadowing of objects, is rather time consuming. Thus, most of the environment mapping algorithms simply ignore this factor and take into account the environment illumination everywhere and in all possible illumination directions.

4.2.1 Mirrored Reflections and Refractions

Let us assume that there are no occlusions, so $v(\mathbf{x}, \omega') = 1$. If the surface is an ideal mirror, then its BRDF allows the reflection just from a single direction, thus the reflected radiance formula simplifies to

$$L^r(\mathbf{x}, \omega) = \int_{\Omega'} L^{\text{env}}(\omega') f_r(\omega', \mathbf{x}, \omega) \cos^+ \theta'_\mathbf{x} v(\mathbf{x}, \omega') \, d\omega' = L^{\text{env}}(\mathbf{R}) F(\mathbf{N} \cdot \mathbf{R}),$$

where $\mathbf{N}$ is the unit surface normal at $\mathbf{x}$, $\mathbf{R}$ is the unit reflection direction of viewing direction ω onto the surface normal, and F is the *Fresnel function*.

In order to compute the reflection of one vector onto another, HLSL introduced intrinsic `reflect`. Environment mapping approaches can be used to simulate not only reflected but also refracted rays, just the direction computation should be changed from the *law of reflection* to the *Snellius–Descartes law of refraction*. The calculation of refraction direction is also supported by HLSL providing the `refract` function for this purpose. This function expects as inputs the incoming direction, the normal vector, and the reciprocal of the *index of refraction*. Intrinsic function `refract` will return a zero vector when *total reflection* occurs, so no refraction direction can be calculated. We should note that tracing a refraction ray on a single level is a simplification since the light is refracted at least twice to go through a refractor. Here we discuss only this simplified case (methods addressing multiple refractions are presented in Chapter 6).

The amount of refracted light can be computed using weighting factor $1 - F$, where F is the Fresnel function (Section 1.1). While the light traverses inside the object, its intensity decreases exponentially according to the *extinction coefficient*. For metals, the extinction coefficient is large, so the refracted component is completely eliminated (metals can never be transparent). For transparent dielectric materials, on the other hand, we usually assume that the extinction coefficient is negligible, and therefore the light intensity remains constant inside the object.

Putting all these together we can implement the environment mapping shader for an object that both ideally reflects and refracts light (e.g. glass). The Fresnel factor for perpendicular illumination (`Fp`) and the index of refraction (`n`) of the object are uniform parameters. The vertex shader transforms the vertex to clipping space as usual, and also obtains the normal vector and the view direction in world space. These world space vectors will be used to compute the reflected and refracted directions.

```
float3 Fp;      // Fresnel at perpendicular dir.
float  n;       // Index of refraction
float3 EyePos;  // Eye position in world space

void EnvMapVS(
    in  float4 Pos   : POSITION, // Input position in modeling space
    in  float3 Norm  : NORMAL,   // Normal vector in modeling space
    out float4 hPos  : POSITION, // Output position in clipping space
    out float3 wNorm : TEXCOORD0,// Normal vector in world space
    out float3 wView : TEXCOORD1 // View direction in world space
) {
    hPos = mul(Pos, WorldViewProj);
    wNorm = mul(Norm, WorldIT);         // Normal in world space
    wView = EyePos - mul(Pos, World);   // View in world space
}
```

The fragment shader gets the interpolated world space normal and view vectors. First it normalizes them, then reflection and refraction directions are calculated. Note that the `refract` function must be called with the reciprocal of the index of refraction. The environment cube map, called `EnvMap`, is accessed using the reflection and refraction directions to get the illumination from these directions. Finally, the Fresnel function is approximated and used to

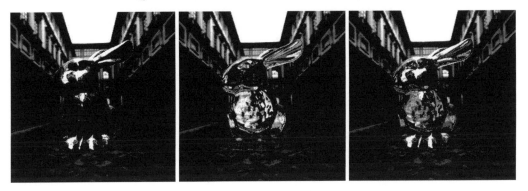

FIGURE 4.9: Environment mapped reflection (left), refraction (middle), and combined reflection and refraction (right).

compute the radiance due to ideal reflection and refraction. Figure 4.9 shows images where the bunny is rendered with this shader.

```
samplerCUBE EnvMap; // environment map

float4 EnvMapPS ( float3 Norm : TEXCOORD0, // World space normal vector
                 float3 View : TEXCOORD1  // World space view vector
               ) : COLOR
{
    float3 N = normalize(Norm);
    float3 V = normalize(View);
    float3 R = reflect(V, N);              // Reflection direction
    float3 T = refract(V, N, 1/n);         // Refraction direction
    // Incoming radiances from the reflection and refraction directions
    float4 Ireflect = texCUBE(EnvMap, R);
    float4 Irefract = texCUBE(EnvMap, T);
    // Approximation of the Fresnel Function
    float cos_theta = -dot(V, N);
    float F = Fp + pow(1-cos_theta, 5.0f) * (1-Fp);
    // Computation of the reflected/refracted illumination
    return F * Ireflect + (1-F) * Irefract;
}
```

4.2.2   Diffuse and Glossy Reflections Without Self-shadowing
Classical environment mapping can also be applied for both glossy and diffuse reflections. However, unlike ideal reflection or refraction, diffuse and glossy reflections do not simplify the integral of the reflected radiance, thus they would require a lot of secondary rays to be traced

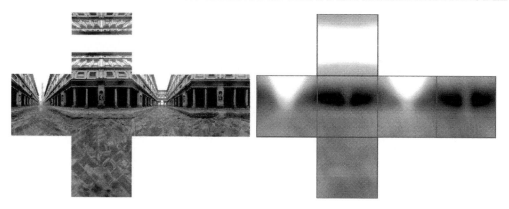

FIGURE 4.10: Original high dynamic range environment map (left) and the diffuse irradiance map (right).

and the integral to be estimated from many samples. Fortunately, if we ignore self occlusions ($v(\mathbf{x}, \omega') = 1$), the time-consuming part of the computation can be moved to a preprocessing phase [111].

Let us consider evaluating the integral of equation (4.2) using a finite number of samples. For example, approximating the reflected radiance with M samples requires the evaluation of the following numerical quadrature (i.e. a sum approximating the integral):

$$L^r(\mathbf{x}, \omega) \approx \sum_{i=1}^{M} L^{\text{env}}(\omega_i') f_r(\omega_i', \mathbf{x}, \omega) \cos^+ \theta_i' \, \Delta\omega_i'.$$

In the case of diffuse objects the BRDF is constant, thus it can be moved out from the sum, and the reflected radiance becomes the product of the diffuse reflectivity and the estimate of the irradiance:

$$L^r(\mathbf{x}, \omega) \approx f_r(\mathbf{x}) \sum_{i=1}^{M} L^{\text{env}}(\omega_i') \cos^+ \theta_i' \, \Delta\omega_i'.$$

The only factor in the irradiance sum that depends on point $\mathbf{x}$ is $\cos^+ \theta_i' = (\mathbf{N_x} \cdot \omega_i)^+$, which is the cosine of the angle between the sample direction and the surface normal of the shaded point. Thus, the irradiance depends just on the normal vector of the shaded surface. We can pre-compute these sums for a sufficient number of normal vectors and store irradiance values

$$I_j = \sum_{i=1}^{N} L^{\text{env}}(\omega_i')(\mathbf{N}_j \cdot \omega_i')^+ \, \Delta\omega_i'.$$

FIGURE 4.11: Diffuse objects illuminated by an environment map.

In order to compute the reflected radiance, we need to determine the normal at the shaded point, look up the corresponding irradiance value, and modulate it with the diffuse BRDF. Irradiance values are stored in an environment map called *diffuse environment map* or *irradiance environment map*. Assuming that each value I_j is stored in a cube map in the direction of the corresponding normal vector $\mathbf{N}_j$, we can determine the illumination of an arbitrarily oriented surface patch during rendering with a single environment map lookup in the normal direction of the surface. In other words, the *query direction* for the environment map lookup will be the surface normal. Assuming cubic environment maps, the pixel shader implementation is as follows:

```
float4 kd; // Diffuse reflection coefficient

float4 DiffuseEnvMapPS(float3 N : TEXCOORD0) : COLOR0 {
    return texCUBE(EnvMap, N) * kd; // radiance = irradiance * BRDF
}
```

In the case of glossy objects, the reflected radiance depends on both the normal direction and the view direction, but this dependence can be described with a single vector. Let us take the original *Phong reflection model* that expresses the reflected radiance as

$$L^r(\mathbf{x}, \mathbf{V}) = L^{\mathrm{env}}(\omega_i')k_s\left((\mathbf{R} \cdot \omega_i')^+\right)^n,$$

where k_s is the glossy reflectivity, n is the shininess of the surface, and $\mathbf{R}$ is *reflection direction* that is the reflection of the viewing direction onto the surface normal. This reflection direction can serve as the query direction in the case of glossy objects, and we can store the following irradiance at reflection direction $\mathbf{R}_j$:

$$I_j = \sum_{i=1}^{M} L^{\mathrm{env}}(\omega_i')\left((\mathbf{R}_j \cdot \omega_i')^+\right)^n \Delta\omega_i'.$$

FIGURE 4.12: Glossy objects illuminated by an environment map.

Having the glossy irradiance map that stores these values, the computation of the glossy reflection is done by the following fragment shader:

```
float4 ks;          // Glossy reflection coefficient
float  shininess;

float4 GlossyEnvMapPS(float3 N : TEXCOORD0,
                      float3 V : TEXCOORD1) : COLOR0
{
    V = normalize(V);
    N = normalize(N);
    float3 R = reflect(V, N);      // Compute the reflection direction
    return texCUBE(EnvMap, R) * ks;  // Reflected radiance
}
```

The pre-calculation of the irradiance environment map is rather computation intensive for both diffuse and glossy objects. In order to obtain a single irradiance value, we have to consider all possible incoming directions and sum up their cosine-weighted incoming intensities. This *convolution* must be performed for each texel of the irradiance environment map. The computation process can be sped up using spherical harmonics [79].

4.2.3   Diffuse and Glossy Reflections With Shadowing

Classical image-based lighting algorithms ignore occlusions and assume that the environment is visible from everywhere. If shadows are also needed, occlusions should also be taken into account, thus shadow mapping should be combined with image-based lighting. In order to estimate the integral of equation (4.2), directional domain Ω' is decomposed to solid angles $\Delta\omega'_i$, $i = 1, \ldots, N$ so that the radiance is roughly uniform within each solid angle, the solid angles are small, and the light flux arriving from them has roughly the same magnitude. If the solid angles are small, then it is enough to test the visibility of the environment map with a single sample in each solid angle:

$$L^r(\mathbf{x}, \omega) = \sum_{i=1}^{N} \int_{\Delta\omega_i'} L^{\text{env}}(\omega') f_r(\omega', \mathbf{x}, \omega) \cos^+ \theta_\mathbf{x}' v(\mathbf{x}, \omega') \, d\omega' \approx \sum_{i=1}^{N} v(\mathbf{x}, \omega_i') \tilde{\Psi}_i^{\text{env}} a(\Delta\omega_i', \omega),$$

where

$$\tilde{\Psi}_i^{\text{env}} = \int_{\Delta\omega_i'} L^{\text{env}}(\omega') \, d\omega'$$

is the *total incoming radiance* from solid angle $\Delta\omega_i'$, and

$$a(\Delta\omega_i', \omega) = \frac{1}{\Delta\omega_i'} \int_{\Delta\omega_i'} f_r(\omega', \mathbf{x}, \omega) \cos^+ \theta_\mathbf{x}' \, d\omega'$$

is the *average reflectivity* from solid angle $\Delta\omega_i'$ to viewing direction ω. In order to use this approximation, the following tasks need to be solved:

1. The directional domain should be decomposed meeting the prescribed requirements, and the total radiance should be obtained in them. Since these computations are independent of the objects and of the viewing direction, we can execute them in the preprocessing phase.

2. The visibility of the environment map in the directions of the centers of the solid angles needs to be determined. Since objects are moving, this calculation is done on the fly. Note that this step is equivalent to shadow computation assuming directional light sources.

3. The average reflectivity values need to be computed and multiplied with the total radiance and the visibility for each solid angle. Since the average reflectivity also depends on the normal vector and viewing direction, we should execute this step as well on the fly.

For a static environment map, the generation of sample directions should be performed at loading time, making it a non-time-critical task. The goal is to make the integral quadrature accurate while keeping the number of samples low. This requirement is met if solid angles $\Delta\omega_i'$ are selected in a way that the environment map radiance is roughly homogeneous in them, and their area is inversely proportional to this radiance. The task is completed by generating random samples with a probability proportional to the radiance of environment map texels, and applying *Lloyd's relaxation* [91] to spread the samples more evenly. A weighted version of the relaxation

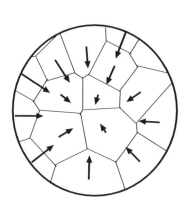

FIGURE 4.13: Radiance values of texels within the Voronoi areas are summed to compute their total power and the final Delaunay grid on high dynamic range image. The centers of Voronoi cells are the sample points.

method is used to preserve the density distribution. As the basic idea of Lloyd's relaxation is to move the sample points to the center of their respective *Voronoi areas*, the relatively expensive computation of the Voronoi mesh is necessary.

A Voronoi cell is defined as the set of points to which a given sample is the closest one. The dual of the Voronoi decomposition is the *Delaunay mesh* of triangles, with the defining property that no circumcircle of any triangle contains any of the other sample points. On the unit sphere of directions, this means that no sample can lie on the outer side of a triangle's plane. A new sample point is inserted according to the *Bowyer–Watson algorithm*:

1. start with a Delaunay mesh that is a convex polyhedron which only has triangle faces, and all vertices are on the unit sphere,

2. delete all triangles for which the new sample direction lies on the outer side of the triangle's plane (where the outer side is defined by the order of the triangle vertices, but for usual meshes the inner side is where the origin is),

3. connect the new sample to the edges of the cavity just created (it will always be a convex cavity in the sense that the resulting mesh will be a legal convex Delaunay polyhedron).

Alternatively, extremely fast algorithms using *Penrose tiling* [102] and *Wavelet importance sampling* have been published to attack the well-distributed importance sampling problem. These methods should be considered if non-static environment maps are to be used. However, if we wish to assign the most accurate light power values to the sampled directions, we have to add up the contributions of those texels that fall into the Voronoi region of the direction.

FIGURE 4.14: Spheres in Florence rendered by taking a single sample in each Voronoi cell, assuming directional light sources (left) and using a better approximation [10] for the average reflectivity over Voronoi cells (right). The shininess of the spheres is 100.

In this case, Lloyd's relaxation means no significant overhead, as rigorously summing the texel contributions takes more time.

Visibility determination

We have decomposed the environment map to solid angles and we want to test visibility with a single sample in each subdomain. The problem is traced back to rendering shadow caused by directional light sources. In order to render shadows effectively, a hardware-supported shadow technique has to be applied. *Shadow maps* [156] generated for every discrete direction are well suited for the purpose.

Lighting using the samples

When the average reflectivity $a(\Delta\omega_i', \omega)$ from solid angle $\Delta\omega_i'$ to viewing direction ω is computed, we have to accept simplifications to make the method real-time. Particularly, the integral of the average reflectivity may be estimated from a single sample associated with the center of the Voronoi region.

However, if the surface is more glossy or the Voronoi cells are large, then the approach results in artifacts. The reflection will be noisy even at points where there is no self occlusion. The problem is that both the albedo and the shadowing factor are estimated with the same, low number of discrete samples. In such cases, the two computations should be separated. We

FIGURE 4.15: Two images of a difficult scene, where only a small fraction of the environment is visible from surface points inside the room. 1000 directional samples are necessary, animation with 5 FPS is possible.

need better approximations for the albedo integral, but use the low number of discrete samples for shadow estimation [10] (Figure 4.14).

Rendering times depend heavily on the number of samples used. Real-time results (30 FPS) are possible for moderately complex scenes with a few objects. However, difficult scenes (Figure 4.15) require more samples for acceptable quality, and therefore only run at interactive frame rates (5 FPS).

CHAPTER 5

Ray Casting on the GPU

In Chapter 1 we introduced the global illumination problem as the solution of the rendering equation. To evaluate the integral numerically, we need samples, and those samples are light paths. In one way or another, all algorithms have to generate light paths connecting the lights with the camera. The most straightforward and general approach is to use *ray casting*, which intersects scene surfaces with a half-line called the *ray* and returns the intersection that is closest to the start of the ray. Classic and successful global illumination methods such as *path tracing* [72], *photon mapping* [69, 68, 70], or *virtual light sources* [78] rely on this operation.

However, these methods could not be applied to GPUs directly. While the GPU outperforms the CPU on ray-intersection calculation because of its parallelism, the CPU has been more efficient at maintaining and traversing the complex data structures needed to trace rays efficiently [59, 150, 116]. Although it is possible to build hardware to support ray-tracing [162], such equipment is not as widely available as GPUs.

Thus, despite the architectural problems, researchers have found various ways to make use of the GPU for ray casting, even if in a limited or sub-optimal way. In this chapter, we review the possibilities, and detail two very different approaches. One of them is the *ray engine* algorithm, which was able to run on the earliest programmable hardware. We also add a trick using cones to make use of the vertex shader and speed up the original brute force approach. The second, more detailed example traverses a complex spatial hierarchy structure on the GPU: the *loose kd-tree*. As this operation requires a stack, the algorithm is designed for Shader Model 4.0 GPUs.

5.1 RAY–TRIANGLE INTERSECTION IN A SHADER

The fundamental operation of ray casting is the *ray–primitive intersection*. Although one advantage of ray-tracing over incremental image synthesis is that it is able to handle various primitives including quadratic and quartic surfaces like spheres or torii, we will only consider triangles. The reason for this is that we are likely to render the same models incrementally, when they are directly visible. Therefore, they must be triangle mesh models, or tessellated to triangles.

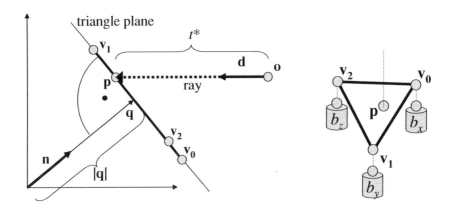

FIGURE 5.1: Nomenclature of the ray–triangle intersection calculation. Barycentric weights (b_x, b_y, b_z) identify the point at the center of mass.

The methods we present can easily be extended to, e.g. quadratic surfaces, or other primitives for which the intersection can be performed in the shader.

In the case of a triangle, the intersection computation consists of finding an intersection point with the triangle's plane, and then a decision of whether the point is within the triangle. See Figure 5.1 for the nomenclature of the intersection test.

In Section 2.1.4 and particularly in equations (2.2) and (2.3), we described a plane by a homogeneous four-tuple, i.e. by four values that can be freely scaled by a nonzero scalar. In order to make the plane definition more compact, now we introduce another representation that needs only three, but not freely scalable values. Let us denote the Cartesian triplet of the nearest point of the plane to the origin by $\mathbf{q}$. Point $\mathbf{q}$ is on the plane, thus it is a place vector of the plane. On the other hand, the direction from the origin to $\mathbf{q}$ is perpendicular to the plane, thus $\mathbf{q}$ is also the normal vector of the plane. Then the equation of the plane is the following:

$$\mathbf{q} \cdot \mathbf{x} = \mathbf{q} \cdot \mathbf{q}. \tag{5.1}$$

Note that this representation is not good if the plane crosses the origin since in this case we would obtain a null vector for the normal. This special case is handled by translating the plane crossing the origin a little in the direction of its normal vector before obtaining vector $\mathbf{q}$.

The parametric ray equation is

$$\mathbf{r}(t) = \mathbf{o} + \mathbf{d}t,$$

where $\mathbf{o}$ is the origin of the ray, $\mathbf{d}$ is the normalized ray direction, and t is the distance along the ray. Substituting the ray equation into the plane equation, we get the following equation for the ray parameter t^* of the intersection:

$$\mathbf{q} \cdot (\mathbf{o} + \mathbf{d}t^*) = \mathbf{q} \cdot \mathbf{q}.$$

From this we can express the ray parameter as

$$t^* = \frac{\mathbf{q} \cdot \mathbf{q} - \mathbf{q} \cdot \mathbf{o}}{\mathbf{q} \cdot \mathbf{d}}.$$

Using the ray equation, the hit point $\mathbf{p}$ is

$$\mathbf{p} = \mathbf{o} + \mathbf{d}t^*.$$

We have to decide whether the point is within the triangle. We prefer methods that also yield the *barycentric coordinates* of the point (Figure 5.1). If the triangle vertex positions are column vectors $\mathbf{v}_0, \mathbf{v}_1, \mathbf{v}_2$, the barycentric coordinate vector $\mathbf{b} = (b_x, b_y, b_z)$ of point $\mathbf{x}$ is defined to fulfil the following equation:

$$\mathbf{v}_0 b_x + \mathbf{v}_1 b_y + \mathbf{v}_2 b_z = [\mathbf{v}_0, \mathbf{v}_1, \mathbf{v}_2] \cdot \mathbf{b} = \mathbf{x}.$$

This means that the barycentric coordinate elements are weights assigned to the triangle vertices. The linear combination of the vertex positions with these weights must give point $\mathbf{x}$. If all three barycentric weights are positive, then the point is within the triangle. If we find the barycentric coordinates for the hit point $\mathbf{p}$, we are not only able to tell if it is within the triangle, but the weights can be used to interpolate normals or texture coordinates given at the vertices. Using the above definition, $\mathbf{b}$ can be expressed as

$$\mathbf{b} = [\mathbf{v}_0, \mathbf{v}_1, \mathbf{v}_2]^{-1} \cdot \mathbf{p},$$

using the inverse of the 3×3 matrix assembled from the vertex coordinates. Let us call the inverse of the vertex position matrix the *IVM*.

Thus, in order to evaluate intersection, we need $\mathbf{q}$ and the IVM. As all three vertices are on the plane, using equation (5.1) and interpreting $\mathbf{q}$ as a row vector, we obtain

$$\mathbf{q} \cdot [\mathbf{v}_0, \mathbf{v}_1, \mathbf{v}_2] = [\mathbf{q} \cdot \mathbf{q}, \mathbf{q} \cdot \mathbf{q}, \mathbf{q} \cdot \mathbf{q}] = |\mathbf{q}|^2 [1, 1, 1]. \tag{5.2}$$

Dividing by $|\mathbf{q}|^2$, we get

$$\frac{\mathbf{q}}{|\mathbf{q}|^2} \cdot [\mathbf{v}_0, \mathbf{v}_1, \mathbf{v}_2] = [1, 1, 1]. \tag{5.3}$$

Let $\mathbf{q}'$ be

$$\mathbf{q}' = \frac{\mathbf{q}}{|\mathbf{q}|^2},$$

and let us multiply equation (5.3) with the IVM from the right:

$$\mathbf{q}' = [1, 1, 1] \cdot [\mathbf{v}_0, \mathbf{v}_1, \mathbf{v}_2]^{-1}.$$

$\mathbf{q}'$ has the same direction as $\mathbf{q}$ and the reciprocal of its length. In turn, $\mathbf{q}$ can be expressed as

$$\mathbf{q} = \frac{\mathbf{q}'}{\mathbf{q}' \cdot \mathbf{q}'}.$$

This means that $\mathbf{q}$ can easily be computed from the IVM. Therefore, we need only the IVM to perform intersection computations. With nine floating-point values, it is a minimal representation for a triangle, with a footprint equal to the vertex positions themselves.

An HLSL function that performs the above computation is presented below. Parameter tBest contains the smallest positive t^* value of previously evaluated intersections, and returns the same taking the current triangle into account. If a hit is found, true is returned in hit, the barycentric weights in b and the hit point in p.

```
void intersect(
    in float3 o, in float3 d,                      // Ray origin and direction
    in float3 ivm0, in float3 ivm1, in float3 ivm2, // Rows of the IVM
    inout float tBest,                             // Nearest hit
    out bool hit,                                  // Has hit occured?
    out float3 b, out float3 p)   // Barycentric coordinates and the hit point
{
    float3 qPrime = ivm0 + ivm1 + ivm2;
    float q = qPrime / dot(qPrime, qPrime);
    hit = false; b = 0; p = 0;
    float tStar = (dot(q, q) - dot(o, q)) / dot(d, q);
    if(tStar > 0.001 && tStar < tBest) {
        p = o + d * tStar;        // Intersection with the plane of the triangle
        b.x = dot(ivm0, p);       // Compute barycentric coordinates
        b.y = dot(ivm1, p);
        b.z = dot(ivm2, p);
        if( all(b > 0) ) {        // If all barycentric coordinates are positive, then
            tBest = tStar;        // the intersection is inside the triangle.
            hit = true;
        }
    }
}
```

This function may be invoked from any shader. Inputs for the intersection consist of ray data and triangle data. These might be stored in the vertex buffer or in textures. Today, almost every combination is feasible. For instance, it is a valid idea to store ray data in a vertex buffer and perform intersection testing on triangles read from textures. In another approach, when

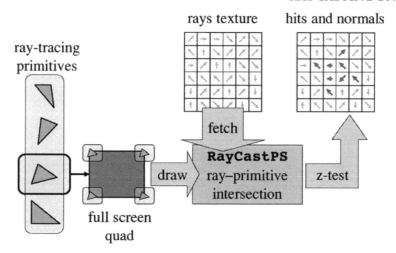

FIGURE 5.2: Rendering pass implementing the ray engine.

rendering a mirror surface, we are able to spawn reflection rays in the fragment shader and test them against all triangles in a single loop.

However, a few years ago shader programs had more limitation, they were not executed dynamically, they could not access textures randomly and repeatedly, and, most importantly, they had no more than a dozen of registers for local memory. In the following section, we review a method that managed to invoke a fragment shader for every ray–triangle pair, and selected the nearest solution for a ray using the depth test hardware.

5.2   THE RAY ENGINE

The *ray engine* [19] is based on the recognition that ray casting is a crossbar on rays and triangles, while rasterization is a crossbar on pixels and primitives. For a set of rays given in textures, and a list of triangles, the ray engine computes all possible ray–triangle intersections on the GPU.

Let us reiterate the working mechanism in current GPU terminology. Figure 5.2 depicts the rendering pass realizing the algorithm. Every pixel of the render target is associated with a ray. The origin and direction of rays to be traced are stored in textures that have the same dimensions as the render target. One after the other, a single triangle is taken, and a viewport sized quad (also called full-screen quad) is rendered for it, with the triangle data attached to the quad vertices. Thus, fragment shaders for every pixel will receive the triangle data, and can also access the ray data via texture reads. The ray–triangle intersection calculation can be performed in the shader, which also outputs the hit distance to the DEPTH register. Using this value, z-buffer hardware performs the depth test to verify that no closer intersection has been found yet. If the result passes the test, it is written to the render target and the depth buffer is updated. This way every pixel will hold the information about the nearest intersection between the scene triangles and the ray associated with the pixel. The pitfall of the ray engine is that it implements the naive ray casting algorithm of testing every ray against every primitive.

Please note that the ray engine is not limited to triangles, but we stick to the term to avoid confusion between ray casting primitives and scan conversion primitives. The ray engine is not a full ray-tracing solution as it does not define how the rays are to be generated, or how the results are used. As originally presented, it is a co-processor to CPU-based ray-tracing, where the CPU delegates some rays and reads back the results. This concept has been used in cinematic rendering software *Gelato*. However, this approach requires frequent communication of rays and results from the GPU to the CPU.

A fragment shader for the ray engine receives the IVM representation of the triangle (as described in Section 5.1) from the vertex stream in TEXCOORD registers. The data assigned to the vertices of the triangle should be interpolated to find the value at the hit point. As an example here, vertex normals are stored in additional TEXCOORD slots. We find the interpolated surface normal by weighting triangle vertex normals with the barycentric coordinates. If we also wish to add texturing, we need to encode and interpolate the texture coordinates of the triangle vertices in a similar manner.

Ray data is read from textures bound to rayOriginSampler and rayDirSampler. The texture address is computed from the pixel coordinate provided in pixel shader input register VPOS. The distance to the hit point, scaled to be between 0 and 1, is written to the DEPTH register. The actual output to render targets depends on the application, it might be a triangle ID with the barycentric coordinates, a refracted ray, or shaded color. In the following example, we exploit the *multiple render targets* feature of the GPU and simultaneously output the surface point position and the shading normal, which can be used in a later shading pass.

```
void RayCastPS(
    in float3  Ivm0    : TEXCOORD0,  // Rows of the IVM
    in float3  Ivm1    : TEXCOORD1,
    in float3  Ivm2    : TEXCOORD2,
    in float3  Normal0 : TEXCOORD3,  // Vertex normals
    in float3  Normal1 : TEXCOORD4,
    in float3  Normal2 : TEXCOORD5,
    in float2  vPos    : VPOS,       // Screen position
    out float4 oHit    : COLOR0,     // Output hit position
    out float4 oNormal : COLOR1,     // Normal at the hit position
    out float1 oDepth  : DEPTH       // Ray parameter at the hit position
) {
```

```
float2 texPos = vPos.xy;            // Pixel to texture coordinates
texPos /= RAY_TEXTURE_SIZE;
float3 o = tex2D(rayOriginSampler, texPos); // Origin
float3 d = tex2D(rayDirSampler, texPos);    // Direction
float t = MAX_DEPTH;
bool hit = false;
float3 b, p;
intersect(o, d, Ivm0, Ivm1, Ivm2, t, hit, b, p);
if ( hit ) {
    oDepth = t / MAX_RAY_DEPTH;  // scale into [0,1]
    float3 normalAtHit = Normal0 * b.x + Normal1 * b.y + Normal2 * b.z;
    normalAtHit = normalize(normalAtHit);
    oHit = float4(p, 1);
    oNormal = float4(normalAtHit, 0);
} else {            // There is no hit
    oDepth = 1; // set to ``infinity''
}
}
```

5.3 ACCELERATION HIERARCHY BUILT ON RAYS

CPU-based acceleration schemes are spatial object hierarchies. The basic approach is that, for a ray, we try to exclude as many objects as possible from intersection testing. This cannot be done in the ray engine architecture, as it follows a per primitive processing scheme instead of the per ray philosophy. Therefore, we also have to apply an acceleration hierarchy the other way round, not on the objects, but on the rays.

In typical applications, real-time ray casting augments scan conversion image synthesis where recursive ray-tracing from the eye point or from a light sample point is necessary. In both scenarios, the primary ray impact points are determined by rendering the scene from either the eye or the light. As nearby rays hit similar surfaces, it can be assumed that reflected or refracted rays may also travel in similar directions, albeit with more and more deviation on multiple iterations. If we are able to compute enclosing cones for groups of nearby rays, it is possible to exclude all rays within a group based on a single test against the primitive being processed. This approach fits well to the ray engine. Whenever the data of a primitive is processed, we do not render a quad for it over the entire screen, but invoke the pixel shaders only where an intersection is possible. This is done (as illustrated in Figure 5.3) by splitting the render target into tiles. We render a set of tile quads instead of a full screen one, but make a decision for every tile beforehand whether it should be rendered at all. Instead of small quads, one can use point primitives, described by a single vertex. This eliminates the fourfold overhead of processing the same vertex data for all quad vertices. The high level test of whether a tile may include valid intersections can be performed in the vertex shader.

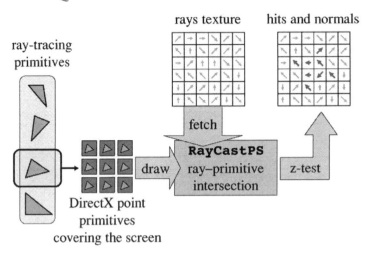

FIGURE 5.3: Point primitives are rendered instead of full screen quads, to decompose the array of rays into tiles.

To be able to perform the preliminary test, an enclosing infinite cone must be computed for rays grouped in the same tile. If we test whether it intersects a triangle, we can skip rendering the tile when it cannot contain an intersection with any of the rays. If the ray texture itself is filled by the GPU, the computation of ray-enclosing cones should also be performed in a rendering pass, computing data to a texture.

Figure 5.4 shows how the hierarchical ray engine pass proceeds. For all the tiles, a vertex buffer with all triangles is drawn. For a faster test, the enclosing spheres of triangles are also included in the vertex buffer. The tile position and the ray-enclosing object description for the current tile are uniform parameters to the vertex shader. Based on the intersection test between the current triangle's and the tile's enclosing object, the vertex shader either transforms the vertex out of view, or moves it to the desired tile position. The pixel shader performs the classic ray engine ray–triangle intersection test.

5.3.1   Implementation of Recursive Ray-tracing Using the Ray Engine

Figure 5.5 depicts the data flow of an application tracing refracted rays. The ray engine passes are iterated to render consecutive refractions of rays. The pixel shader performing the intersection tests outputs the refracted ray. That is, the ray defining the next segment of the refraction path is written to the render target. Then these results are copied back to the ray texture, and serve as input for the next iteration. Those pixels in which the path has already terminated must not be processed. At the beginning of every iteration, an enclosing cone of rays is built for every

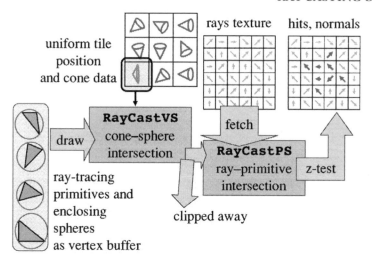

FIGURE 5.4: The rendering pass implementing the hierarchical ray engine. For every tile, the vertex buffer containing triangle and enclosing sphere data is rendered. The vertex shader discards the point primitive if the encoded triangle's circumsphere does not intersect the cone of the tile.

tile, and stored in a texture. Data from this texture is used in ray-casting vertex shaders to carry out preliminary intersection tests. Note that the cones have to be reconstructed for every new generation of rays, before the pass computing the nearest intersections.

The steps of the complete algorithm rendering a frame in an interactive animation sequence can be listed as follows:

1. Generate the primary ray array. This is done by rendering the scene with a shader that renders the scene geometry with standard transformations but outputs reflected or refracted, i.e. secondary rays.

2. Compute the enclosing cones for tiles of rays.

3. For every tile, draw the vertex buffer encoding scene triangles, using the ray engine shader, rendering secondary rays where an intersection is found.

4. Copy valid secondary rays back to the ray texture.

5. Repeat from step 2 until desired recursion depth is reached.

6. Draw a full-screen quad, using the secondary exiting rays in the ray texture e.g. to address an environment map. If we trace refractive rays, then this renders the refractive objects with environment mapping.

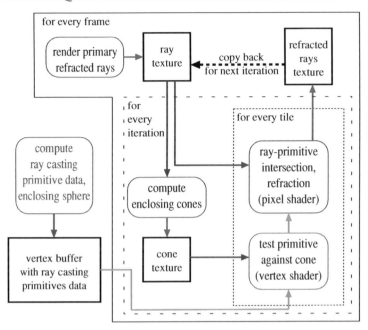

FIGURE 5.5: Block diagram of the recursive ray-tracing algorithm. Only the initial construction of the vertex buffer is performed on the CPU.

5.3.2   Ray Casting

In the steps before ray casting, we have rendered an array of rays to be traced, specified by the origin and direction in world space. We have built a texture of enclosing cones, also in world space, corresponding to rays of tiles. Now, in the ray casting pass, we have to test these against enclosing spheres of triangles encoded in vertices, given in modeling space. We transform the cones and rays to modeling space for the intersection computation, and transform the results back if necessary. The algorithm for the ray–triangle intersection was described in Section 5.1.

The cone intersects the sphere if

$$\varphi > \arccos[(\mathbf{v} - \mathbf{a}) \cdot \mathbf{x}] - \arcsin[r/|\mathbf{v} - \mathbf{a}|],$$

where $\mathbf{a}$ is the apex, $\mathbf{x}$ is the direction, and φ is the half opening angle of the cone, $\mathbf{v}$ is the center of the sphere, and r is its radius. The vertex shader has to test for this, and pass all the information necessary for the ray intersection test to the pixel shader. The cone is transformed using the inverse `World` transform from world to model space. Note that the parameters necessary for the ray–triangle intersection, which the vertex shader simply passes on to the pixel shader have been omitted from this listing for clarity.

```
void RayCastVS(
    in float4 SphereCentre : POSITION,
    in float4 SphereRadius : TEXCOORD0,
    uniform float3 a;                       // Cone apex
    uniform float3 x;                       // Cone direction
    uniform float3 phi;                     // Cone opening
    uniform float3 TilePos;
    out float4 hPos        : POSITION       // Output in clipping space
) {
    x = mul(WorldIT, float4(x, 0));         // Transform cone to model space
    a = mul(WorldIT, float4(a, 1));
    float3 sfc = SphereCentre - a;
    float lsfc = length(sfc);
    float deviation = acos(dot(x, sfc) / lsfc);
    float sphereAngle = asin(SphereRadius / lsfc);
    // Clip away if non-intersecting
    if(phi + sphereAngle < deviation) hPos = float4(2, 0, 0, 1);
}
```

The pixel shader performs the intersection test on ray data read from the textures. The ray is transformed into model space, and the refracted ray origin and direction are transformed back using the World transformation.

```
void RayCastPS(
    in float3  Ivm0    : TEXCOORD0,     // Rows of the IVM
    in float3  Ivm1    : TEXCOORD1,
    in float3  Ivm2    : TEXCOORD2,
    in float3  Normal0 : TEXCOORD3,     // Vertex normals
    in float3  Normal1 : TEXCOORD4,
    in float3  Normal2 : TEXCOORD5,
    in float2  vPos    : VPOS,          // Pixel coordinates
    out float4 oOrigin : COLOR0,        // Refracted ray origin
    out float4 oDir    : COLOR1,        // Refracted ray direction
    out float1 oDepth  : DEPTH          // Ray parameter of the hit
) {
    float2 texPos = vPos.xy;            // Pixel to texture coordinates
    texPos /= RAY_TEXTURE_SIZE;
    // Fetch ray and transform to model space
    float3 o = tex2D(rayOriginSampler, texPos);
    o = mul(WorldIT, float4(o, 1));
    float3 d = tex2D(rayDirSampler, texPos);
    d = mul(WorldIT, float4(d, 0));

    float t = MAX_DEPTH;                              // Intersection calculation
    bool hit = false;
```

```
    float3 b, p;
    intersect(o, d, Ivm0, Ivm1, Ivm2, t, hit, b, p);
    if (hit) {
        oOrigin = mul(float4(p, 1), World);        // Result to world space
        oDepth = t / MAX_RAY_DEPTH;
        // Normal interpolation
        float3 normalAtHit = Normal0 * b.x + Normal1 * b.y + Normal2 * b.z;
        normalAtHit = normalize(normalAtHit);
        if (dot(normalAtHit, d) > 0) {             // Exiting ray
            normalAtHit = -normalAtHit;
            fr = 1.0 / fr;                         // Index of refraction
        }
        float3 refrDir = refract(d, normalAtHit, fr);
        // When total internal reflection occurs, refract returns 0 vector
        if (dot(refrDir, refrDir) < 0.01) refrDir = reflect(d, normalAtHit);
        oDir = mul(float4(refrDir, 0), World);
    } else {          // no hit
        oDepth  = 1; // set to ''infinity''
    }
}
```

5.4   FULL RAY-TRACERS ON THE GPU USING SPACE PARTITIONING

When full ray-tracing is targeted, the brute force intersection test for every triangle and ray is not feasible anymore. The considerable amount of research results obtained in the CPU-based

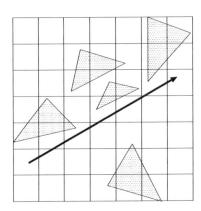

FIGURE 5.6: A uniform grid that partitions the bounding box of the scene into cells of uniform size. In each cell objects whose bounding box overlaps with the cell are registered.

FIGURE 5.7: Reflective objects rendered by Purcell's method at 10 FPS on 512 × 512 resolution using an NV6800 GPU.

ray-tracing acceleration [1, 6, 7, 42, 45, 101, 151, 60] should also migrate to GPU ray-tracing. However, when doing so, the special features of the GPU should also be taken into account.

5.4.1   Uniform Grid

Purcell et al. [109] used a uniform 3D grid partition [42, 8] of the scene geometry (Figure 5.6) which was cache-coherent and accessed in constant-time. Unfortunately, this scheme resulted in a large number of ray steps through empty space.

A uniform space subdivision scheme requires storing the list of referenced triangles for each cell. The cells of a uniform grid are also called *voxels*. To increase the speed of the grid traversal, the uniform grid holds a pre-computed distance transform called a *proximity cloud* [74] which allows the traversal kernel to skip large empty regions of space. The scene representation is stored in an RGB 3D texture, where each grid cell is encoded in one texel. The R component is a pointer to the first referenced triangle (in the voxel) in the list of triangles, G holds the number of the referenced objects, while the value in the third (B) channel gives the distance in voxels to the closest cell containing triangles, or zero for a non-empty voxel. This distance is used for skipping empty cells during grid traversal. In the list of referenced triangles, each texel holds a pointer to actual triangle data. After these two levels of indirection, triangle data itself is stored in two other textures.

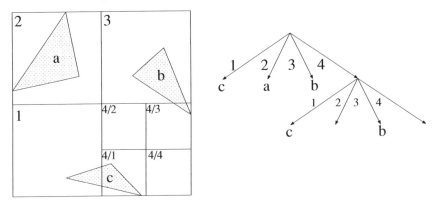

FIGURE 5.8: A quadtree partitioning the plane, whose three-dimensional version is the octree. The tree is constructed by halving the cells along all coordinate axes until a cell contains "just a few" objects, or the cell size gets smaller than a threshold. Objects are registered in the leaves of the tree.

5.4.2   Octree

For *octrees* [44, 8], the cells can be bigger than the minimum cell (Figure 5.8), which saves empty space traversal time. The method has been transported to GPU by Meseth et al. [93]. The hierarchical data structure is mapped to a texture, where values are interpreted as pointers. Chasing those pointers is realized using dependent texture reads. The next cell along a ray is always identified by propagating the exit point from the current cell along the ray, and performing a top–down containment test from the root of the tree. In order to avoid processing lists of triangles in cells, every cell only contains a single volumetric approximation of visual properties within the cell. Thus, scene geometry needs to be transformed into a volumetric representation. Although the method makes use of spatial subdivision, it is in fact a clever approximate ray-tracing scheme over a discretely sampled scene representation.

5.4.3   Hierarchical Bounding Boxes

Geometry images can also be used for the fast bounding box hierarchy [145] construction as has been pointed out in [21]. Here a threaded bounding box hierarchy was used which does not rely on conditional execution (another feature poorly supported by the GPU) to determine the next node in a traversal. The method threads a bounding box hierarchy with a pair of pointers per node, indicating the next node to consider given that the ray either intersects or misses the node's bounding box. These threads allow the GPU to efficiently stream through the hierarchy without maintaining a stack.

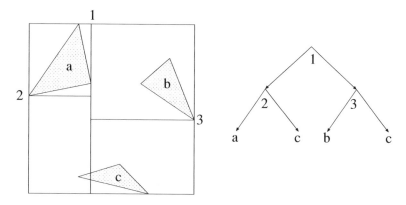

FIGURE 5.9: A kd-tree. Cells containing "many" objects are recursively subdivided into two cells with a plane that is perpendicular to one of the coordinate axes. Note that a triangle may belong to more than one cell.

5.4.4   Kd-Tree

The *kd-tree* is a Binary Space Partitioning (BSP) tree with axis-aligned splitting planes (Figures 5.9 and 5.10). As a space subdivision scheme, it can be used for accelerating ray casting. The use of axis-aligned splitting planes has several advantages. Most importantly, it makes intersection tests inexpensive and allows for low memory usage of the tree representation. The interior nodes of the tree have to store the orientation and the position of the splitting plane. For the leaves, we need to have the list of objects intersecting the corresponding volume. For ray casting, we have to traverse the data structure to identify the leaf cells along the ray, and compute intersection only with objects of those cells.

Among the popular acceleration schemes, the kd-tree is regarded as the most flexible, and thus, generally the most effective one. However, it requires long construction time due to the fact that the optimal axis-aligned cutting plane is always chosen from all possibilities according to a cost estimate. The outline of the building algorithm, starting with a single cell holding all triangles, is the following:

1. Make three copies of the triangle list. Sort them according to their x, y, and z extrema, respectively (at this point, let us assume that the ordering is well defined).

2. Iterate through all arrays, and at triangle extrema, evaluate the predicted traversal cost for the case that the cell is cut into two smaller cells with an axis-aligned plane there. The usual cost estimate is the surface area of the cell multiplied by the number of triangles within: this is called *surface area heuristics* [59]. Keep the best cutting plane position and orientation.

3. If the best cut is better then keeping the original cell, make two new cells, sort triangles into them, and execute the algorithm recursively. The cutting plane data and the references to the children must be written to the tree data structure. If there are no triangles or the cost estimate prefers not to divide the cell, make a leaf. A reference to the list of triangles within the cell must be written to the tree data structure.

Traversal is also a recursive algorithm [59, 131]. We always maintain the ray segment of interest, so that we can exclude branches that the ray does not intersect. For a leaf, we compute intersections with all triangles, keeping only the nearest of those within the ray segment. For a branching node, we identify the *near child cell* and the *far child cell*, that is, the ones the ray visits first and second. We compute the ray segments within the two child cells, traverse the near node first, and traverse the far node only if no valid intersection has been found. It is possible that the ray intersects only one of the children, in which case the other subtree needs not to be processed.

Foley and Sugerman [41] implemented the *kd-tree* acceleration hierarchy on the GPU. Their algorithm is not optimal in algorithmic sense [59, 138, 61], but eliminated the need of a stack, which is very limited on the GPU [36]. While this decomposition showed that GPU hierarchy traversal is feasible, it achieved only 10% of the performance of comparable CPU implementations. The method is not directly targeted on real-time applications, and it is ill-suited for highly dynamic scenes because of the construction cost of the kd-tree. They implemented two variations on a stackless traversal of a kd-tree: kd-restart, which iterates down the kd-tree, and kd-backtrack, which finds the parent of the next traversal node by following parent pointers up the tree, comparing current traversal progress with per-node bounding boxes.

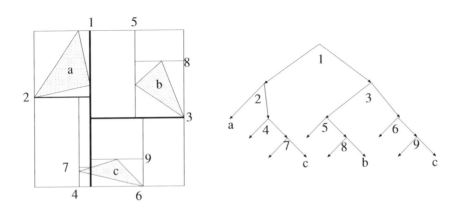

FIGURE 5.10: A kd-tree cutting away empty spaces. Note that empty leaves are possible.

The main caveat of the kd-tree is that the ordering of triangles is not well defined. Triangle bounding boxes typically overlap, and it is not always possible or desirable to only find cutting planes that do not intersect any triangles. This also means triangles may belong to both cells, and the triangle lists for the two leaves together may be longer than the original list, complicating and slowing down the building process. Even more painfully, during traversal, these triangles will be intersected more than once. They also extend out of their cells, so extra tests are necessary to verify that a hit actually happened within the processed cell. If it did not, another intersection in another cell might be closer.

5.5   LOOSE KD-TREE TRAVERSAL IN A SINGLE SHADER

Spatial ray-casting acceleration hierarchies can be constructed on the CPU and stored in textures for the GPU. With a series of dependent texture reads within a dynamically executed loop, they can also be traversed. The key problem is keeping track of parts of the hierarchy that have not been visited yet. For a tree, traversal is usually defined recursively, in the fashion of "traverse left branch, then traverse right branch". Recursive execution is not supported in shaders, so we use iteration instead. To remember what subtrees are to be traced later, we need stack memory. Shader Model 4.0 GPUs have a register file of 4096 units that can be used for this purpose, eliminating the last obstacle from implementing full tree traversal in a shader. Therefore, the implementation presented here is going to differ from other pieces of code within this book in that it uses Shader Model 4.0 compatible shaders and Direct3D 10 features.

While GPUs are still built for coherent execution, and do not excel at conditional statements and dynamic loops, this is now a mere performance optimization issue dwarfed by the available computing power.

Woop et al. [161] have proposed *bounding kd-trees* for ray casting as hybrid spatial indexing hierarchies, combining kd-trees and bounding volume hierarchies. They proposed a specialized hardware architecture for ray-tracing dynamic scenes, emphasizing how bounding kd-trees can be updated quickly. We can use a very similar scheme for GPU ray-tracing, without the need of custom hardware. We will refer to this scheme as the *loose kd-tree*, because it applies the same idea as the loose octree [144].

The loose kd-tree elegantly solves the problem of duplicating triangles and out-of-cell intersections. Instead of cutting the cell into two non-overlapping cells sharing triangles, we decompose it to two possibly overlapping cells, where every triangle belongs strictly to one cell, namely the one that completely contains it (see Figures 5.11 and 5.12). The price is that we have to store the position for two cutting planes: the maximum of the left cell and the minimum of the right cell. Using two 16-bit floats, it is still only four bytes per node. Also, during traversal, we will need to compute intersections with both planes, but on a GPU with

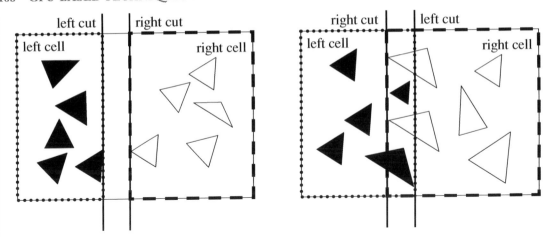

FIGURE 5.11: Two cell decompositions in a loose kd-tree. There is either an empty region or a shared region. Dotted triangles belong to the left cell. Contrary to classic kd-trees, triangles never extend outside their cell.

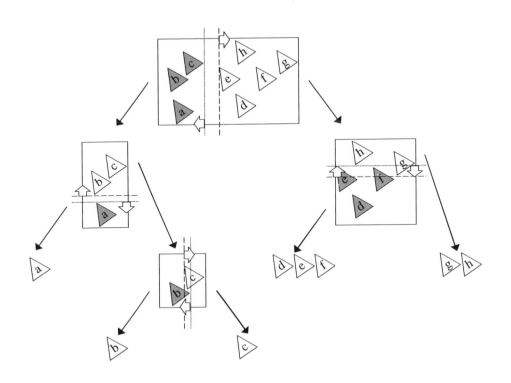

FIGURE 5.12: Construction of a loose kd-tree.

four-channel ALUs this comes for free. The slightly increased storage cost of tree nodes is compensated by the fact that triangles are never duplicated. Furthermore, a loose kd-tree is less sensitive to number representation inaccuracy: the pathologic case of triangles coincident with a cutting plane is easy to resolve.

5.5.1 GPU Representation of the Loose kd-tree

With any type of tree data structure, we should be able to find the child nodes of a parent node. For a balanced binary tree, this is easy to do if the nodes are laid out in an array level by level (the *compact representation*). If the root node has index 0, the child nodes of node n will be $2n + 1$ and $2n + 2$. Unfortunately, neither a kd-tree nor a loose kd-tree is usually balanced. Using pointers for non-leaf nodes to identify children could easily double the memory requirement, and also the amount of data to be read for traversal. A combined method has to be found, which lays out portions of the tree with the above indexing scheme, but maintains pointers for branches exceeding those blocks. Such a scheme has been described in [131]. In our GPU implementation blocks will be rows of a texture (see Figure 5.13). A node is thus identified by a two-element index. Within the rows, children are found using compact indexing. For branching nodes on the second half of the texture row, where indices so computed would exceed the block, we store a pointer. The node indexed by this pointer holds the real data, and its children can also be identified by compact indexing.

The data structure consists of three arrays (in Figure 5.13 the node and flag textures are unified), all of which are bound to the shader as textures.

The leaf texture contains the IVMs for all triangles, in such a way that triangles in the same leaf are next to each other in a continuous block.

The node texture is the main tree structure. Every element is four bytes wide, but can be of three different types. A *leaf node* represents a cell, storing the triangle count and the starting index of the triangle list within the leaf texture. A *cut node* contains two floating-point positions for the left and the right cutting plane position. A *pointer node* contains an index redirecting to another node within the node texture.

The flag texture is the same size as the node texture, and contains 2-bit values specifying how the node in the node texture should be interpreted. The value 3 means a leaf, values 0–2 identify cuts on the x-, y-, and z-axis. A node is a pointer node if it is not a leaf node, but it is on the last level of a node block (or the right half of the texture).

If there is an original model mesh geometry defined by a vertex and an index buffer, then the index buffer is also sorted to have the same ordering as the leaf texture. After the nearest hit

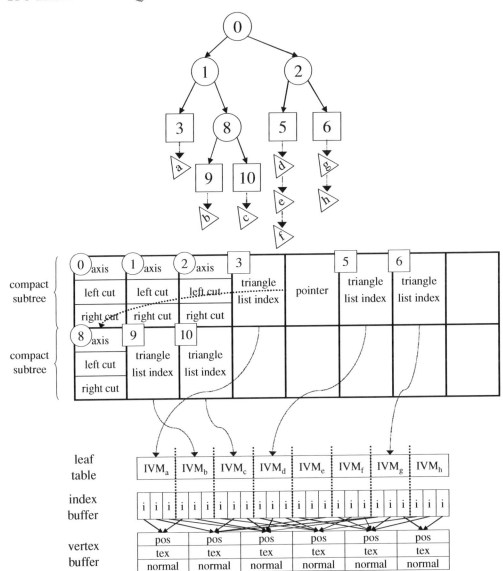

FIGURE 5.13: An example loose kd-tree and its GPU representation.

has been found, texture coordinates or normals of the intersected triangle's vertices can thus be read from the vertex buffer. This makes texturing and smooth shading possible.

The four bytes of a node are divided into two 16-bit integers (triangle count and starting index) for leaf nodes and two 16-bit floats (cut plane positions) for cut nodes. In the shader, this polymorphism is handled by binding the same untyped texture resource of

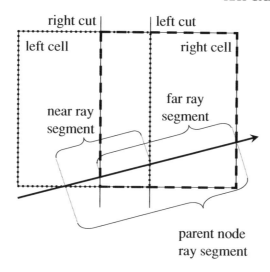

FIGURE 5.14: Traversal of a loose kd-tree branch node. We must compute the distances where the ray pierces the cutting planes to find the child ray segments. If the ray direction along the cut axis is negative, the near and far cells are reversed.

format `R16G16Typeless` using two typed views of format `R16G16UnsignedInteger` and `R16G16Float`. This means that in the shader they are seen as two different textures, even though they occupy the same memory. Note that this feature of untyped resources and resource views is only available in Direct3D 10. Binding the two views to different samplers, we are able to access the same data with different interpretations in a shader. Without this, we would have to encode floats in an integer texture (or vice versa) and decode them in the shader.

5.5.2 GPU Traversal of the Loose kd-tree

The recursive traversal algorithm for the loose kd-tree is implemented as an iterative procedure using a stack. The first node to be processed is the root node. If the node is a pointer node, we de-reference it first to get the corresponding cut node. If the node is a cut node, we compute the intersection distances to the cutting planes along the ray. Using these, the extrema of ray segments within the two child cells are obtained (Figure 5.14). According to the ray direction, the two are labeled the near and the far child. We push the node index of the far cell and the far ray segment onto the stack, if it has positive length. The near child node and near ray segment are processed in the next iteration.

The recursion is terminated in leaf nodes. If the processed node is a leaf, we update the best hit found so far by processing all triangles in the cell. Then the next node to be processed and the ray segment are popped from the top of the stack. The segment maximum is decreased to the distance of the best hit found so far. If the segment does not have positive length, there is nothing to be traced, and the next node can be popped from stack starting a new iteration. If the stack becomes empty, the traversal of the ray is finished.

The shader function processing a single triangle reads the triangle data from the leaf texture, and invokes the intersection function described in Section 5.1. Parameter i identifies a triangle: it is the texture coordinate where the first row of the IVM is stored. After every texture fetch operation, this is advanced to the next texel. Note that we operate directly on textures using the Load method, without samplers, and texture coordinates are integer texel indices. The updated depth, barycentric coordinates and triangle ID for the nearest hit found so far are passed in inout parameters and updated when necessary.

```
void processTriangle(
    in float3    o,     // Ray origin
    in float3    d,     // Ray direction
    inout uint2  i,     // Triangle data index
    inout float  tBest, // Ray parameter
    inout float3 bBest, // Barycentric
    inout float3 pBest, // Hit point
    inout uint   iBest  // Best hit triangle
) {
    // Fetch triangle data
    float3 ivm0 = leafTexture.Load(uint3(i, 0));  i.x++;
    float3 ivm1 = leafTexture.Load(uint3(i, 0));  i.x++;
    float3 ivm2 = leafTexture.Load(uint3(i, 0));  i.x++;
    bool hit;
    float3 b, p;
    intersect( o, d, ivm0, ivm1, ivm2, tBest, hit, b, p);
    if (hit) {
        bBest = b; pBest = p; iBest = i;
    }
}
```

One of the fundamental operations is finding the children of a branch node. In our data structure, because of the compact layout, child nodes are always next to each other. Nodes are identified by two integers, the block index and the node index within the block, which are also their texture coordinates in the node and flag textures. The children of a parent node within a block are found using the classic compact indexing arithmetics. If they exceeded the compact block, then the parent node would in fact be a pointer node, which first had to be de-referenced to get the actual cut node. A pointer node never points to another pointer node, so after this the child index can safely be computed.

```
uint2 getChild(
    inout uint2 parent, // Parent node index
    inout uint flags    // Parent node flags
) {
    // Index of left child within compact block
    uint2 child = uint2(parent.x, (parent.y << 1) + 1);

    // If child is out of block, then parent is a pointer node.
    if (child.y >= nNodesPerBlock) {
        // Follow pointer
        parent = asuint(nodeTexture.Load(uint3(parent.yx, 0)));
        flags =  asuint(flagTexture.Load(uint3(parent.yx, 0)));
        // Index of left child
        child = uint2(parent.x, (parent.y << 1) + 1);
    }
    return child;
}
```

The fragment shader uses the above functions to implement the loose kd-tree traversal algorithm. We set up the stack to contain the root node. The while loop pops the next node/ray segment combination from the traversal stack. For a cut node the children are found, and the ray segments for the child cells are computed. If the ray traverses both cells, one of them is pushed to the stack. The loop goes on with the next node. When we reach a leaf, we perform intersection testing an all contained triangles. The node texture is accessed through two variables: nodeTexture interprets texels as pairs of unsigned integers, while cutTexture assumes two 16-bit floating point numbers. Through which a node must be accessed is defined by values in flagTexture.

```
float3 units[3] = {float3(1, 0, 0), float3(0, 1, 0), float3(0, 0, 1)};

float4 RayCastPS(
    in float3 wPos : TEXCOORD0   // Ray target
    ) : SV_TARGET                // Replaces POSITION in Shader Model 4.0
{
    float3 o = EyePos;           // Set up eye rays
    float3 d = normalize(wPos.xyz - EyePos.xyz);
    float3 invd = 1.0 / d;       // Avoid divisions later
    float4 stack[36];            // Stack big enough for the tree
    float2 segment = float2(0.0, 100000.0); // Ray segment: t=(tMin, tMax).
    uint2 node = 0;              // Root: block 0, node 0
```

```
uint stackPointer = 1;      // Stack with root node
traversalStack[0] = float4(segment, asfloat(node));

float tBest = 1000000.0;  // Max distance
float3 bBest = 0;         // Best hit's barycentric
uint iBest = 0;           // Best hit's triangle index
while(stackPointer > 0) {
    // Pop (node, segment) from stack
    float4 nextNode = traversalStack[--stackPointer];
    node = asuint(nextNode.zw);
    segment = nextNode.xy;
    segment.y = min(tBest, segment.y); // Clamp tMax to tBest
    // Determine node type
    uint nodeFlags = asuint(flagTexture.Load(uint3(node.yx, 0)));
    while(nodeFlags != 3) {
        // Trace pointer and get address of left child
        uint2 nearChild = getChild(node, nodeFlags), farChild = nearChild;
        // Compute intersection with cutting planes
        float2 cuts = cutTexture.Load(uint3(node.yx, 0));
        float3 axisMask = units[nodeFlags];
        float oa = dot(rayOrigin, axisMask);     // o[axis]
        float invda = dot(invRayDir, axisMask);  // invd[axis]
        float2 cellCutDepths = (cuts - oa.xx) * invda.xx;
        // Switch left and right if d[axis] < 0
        if(invda > 0.0) farChild.y += 1; // The right child is the far child
        else {
            cellCutDepths.xy = cellCutDepths.yx;
            nearChild.y += 1;              // The right child is the near child
        }
        // Compute ray segments within child cells
        float2 nearSegment = float2(segment.x, min(segment.y, cellCutDepths.x));
        float2 farSegment = float2(max(segment.x, cellCutDepths.y), segment.y);
        // Find next node
        if(nearSegment.x > nearSegment.y) { // No ray segment for near child
            segment = farSegment; node = farChild;
        } else {
            segment = nearSegment; node = nearChild;
            if(farSegment.x < farSegment.y) // Right child has ray segment
                traversalStack[stackPointer++] = float4(farSegment,asfloat(farChild));
        }
        uint nodeFlags = asuint(flagTexture.Load(uint3(node.yx, 0)));
    }
    // Find leaf list start and end
    uint2 leafList = asuint( nodeTexture.Load(uint3(node.yx, 0)) ) * 3;
    uint2 leafTex = uint2(leafList.y % 2048, leafList.y / 2048);
```

FIGURE 5.15: A scene rendered with textures and reflections (15 000 triangles, 10 FPS at 512×512 resolution).

```
        uint leafEnd = leafTex.x + leafList.x;
        while (leafTex.x < leafEnd) processTriangle(o, d, leafTex, tBest, bBest, iBest);
    }
    return float4(bBest.xyz, 0);
}
```

5.5.3 Performance

The above kd-tree implementation achieves interactive frame rates (10–20 million rays per second) for moderately complex scenes (10 000–100 000 triangles), but its performance depends on the coherence of rays to be traced. If shaders in neighboring pixels branch differently during execution, the pipeline coherence quickly degrades, and the algorithm becomes inferior to multi-pass solutions. The quest is on to find an algorithm that executes traversal and intersection coherently in a single shader.

CHAPTER 6

Specular Effects With Rasterization

The computation of specular effects, such as ideal *reflection*, *refraction*, or *caustics*, can be regarded as a process looking for light paths containing more than one scattering point (Figure 6.1). For example, in the case of reflection we first find the point visible from the camera, which is the place of reflection. Then from the *reflection point* we need to find the *reflected point*, and reflect its radiance at the reflection point toward the camera. Refraction is similar but we have to follow the refraction rather than the reflection direction. Note also that even caustics generation is similar with the difference that the search process starts from the light source and not from the camera. It might be mentioned that in all cases the points correspond to the shortest optical path according to the *Fermat principle*.

Searching for a complete light path in a single step is not easy since it requires the processing of the scene multiple times, which is not compatible with the stream processing architecture of the GPU (this architecture assumes that vertices can be processed independently of each other and fragments can be shaded independently).

GPUs trace rays of the same origin very efficiently taking a "photo" from the shared ray origin, and use the fragment shader unit to compute the radiance of the visible points. The photographic process involves the rasterization of the scene geometry and the exploitation of the z-buffer to find the first hits of the rays passing through the pixels. The computation of the reflected radiance at ideal reflectors or refractors requires secondary rays to be traced, which means that the fragment shader should have access to the scene geometry. Since fragment

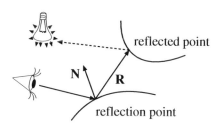

FIGURE 6.1: Specular effects require searching for complete light paths in a single step.

shaders may access global (i.e. uniform) parameters and textures, this requirement can be met if the scene geometry is stored either in uniform parameters or in textures.

A simple approximation of specular reflections is *environment mapping* [16] (see Section 4.2). This approach assumes that the reflected points are very (infinitely) far, and thus the hit points of the rays become independent of the reflection points, i.e. the ray origins. When rays originate at a given object, environment mapping takes images of the environment from the center of the object, then the environment of the object is replaced by a cube, or by a sphere, which is textured by these images (Figure 4.7). When the incoming illumination from a direction is needed, instead of sending a ray we can look up the result from the images constituting the *environment map*. The fundamental problem of environment mapping is that the environment map is the correct representation of the direction-dependent illumination only at its reference point. For other points, accurate results can only be expected if the distance of the point of interest from the reference point is negligible compared to the distance from the surrounding geometry. However, when the object size and the scale of its movements are comparable with the distance from the surrounding surface, errors occur, which create the impression that the object is independent of its illuminating environment (Figure 6.2).

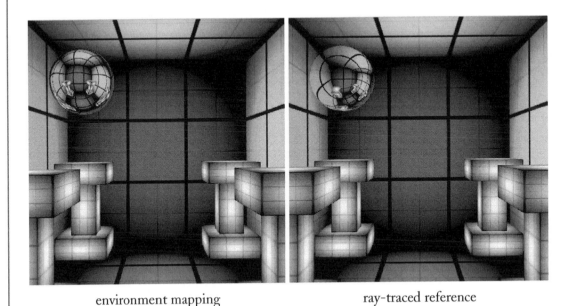

environment mapping   ray-traced reference

FIGURE 6.2: A reflective sphere in a color box rendered by environment mapping (left) and by ray-tracing (right) for comparison. The reference point is in the middle of the color box. Note that the reflection on the sphere is very different from the correct reflection obtained by ray-tracing if the sphere surface is far from the reference point and close to the environment surface (i.e. the box).

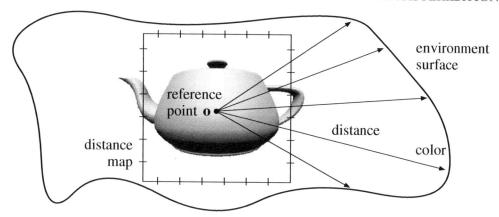

FIGURE 6.3: A distance map with reference point **o**.

To compute the steps of these paths more accurately, rays should be intersected with the scene geometry. Scene objects can be encoded as the set of parameters of the surface equation, similarly to classical ray-tracing [110]. Should the scene geometry be too complex, simple proxy geometries can be used that allow fast approximate intersections [15, 106]. The previous chapter discussed the implementation of the ray-tracing algorithm on the GPU, which uses the pipeline in a way that is not compatible with the classical incremental rendering where primitives are sent down the pipeline one by one. In this chapter, on the other hand, we examine approaches that fit well to the concept of incremental rendering. These approaches are based on the recognition that the scene geometry needed to trace secondary rays can also be represented in a sampled form and stored in textures accessed by the fragment shader. The oldest application of this idea is the *depth map* used by shadow map algorithms [156], but it samples just that part of the geometry which is seen from the light source through a window. To consider all directions, a *cube map* should be used instead of a single depth image. The center of the cube map is called the *reference point*.

If we store world space Euclidean distances in texels, the cube map is called the *distance map* [105, 137] (Figure 6.3). In a distance map a texel, including its address representing a direction and its stored distance, specifies the point visible at this direction without knowing the orientation of the cube map faces. This property is particularly important when we interpolate between two directions, because it relieves us from taking care of cube map face boundaries.

The computation of distance maps is very similar to that of classical environment maps. The only difference is that in addition to the color, the distance is also calculated and stored in the cube map. Since the distance is a nonlinear function of the homogeneous coordinates of the points, correct results can be obtained only by letting the fragment shader compute the distance values. We shall assume that the color information is stored in the r, g, b channels of the cube map texture, while the alpha channel is reserved for floating point distance values.

A major limitation of single distance map-based approaches is that they represent just those surfaces which are visible from the reference point, thus those ray hits which are occluded from the reference point cannot be accurately computed. This limitation is particularly crucial when self reflections or refractions need to be rendered. In this case, more than one distance layer is needed.

In the following section, we discuss an approximate ray-tracing scheme that uses a single-layer distance map and computes single reflections and refractions. Then we shall extend the method for multiple self reflections, which requires more than one layer.

6.1   RAY-TRACING OF DISTANCE MAPS

The basic idea of the ray-tracing process is discussed using the notations of Figure 6.4. Let us assume that center $\mathbf{o}$ of our coordinate system is the *reference point* and we are interested in the illumination of point $\mathbf{x}$ from direction $\mathbf{R}$.

The illuminating point is thus on the ray of equation $\mathbf{r}(d) = \mathbf{x} + \mathbf{R}d$. The accuracy of an arbitrary guess for d can be checked by reading the distance of the environment surface stored with the direction of $\mathbf{l} = \mathbf{x} + \mathbf{R}d$ in the cube map ($|\mathbf{l}'|$) and comparing it with the distance of approximating point $\mathbf{l}$ on the ray ($|\mathbf{l}|$). If $|\mathbf{l}| \approx |\mathbf{l}'|$, then we have found the intersection. If the point on the ray is in front of the surface, that is $|\mathbf{l}| < |\mathbf{l}'|$, the current approximation is *undershooting*. The case when point $\mathbf{l}$ is behind the surface ($|\mathbf{l}| > |\mathbf{l}'|$) is called *overshooting*.

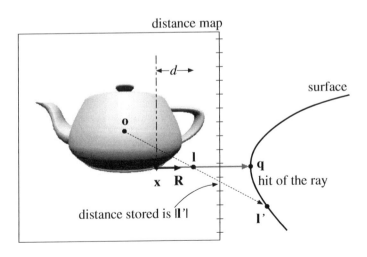

FIGURE 6.4: Tracing a ray from $\mathbf{x}$ at direction $\mathbf{R}$.

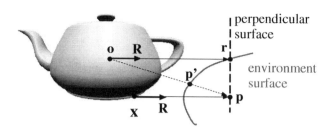

FIGURE 6.5: Approximation of the hit point by **p** assuming that the surface is perpendicular to the ray direction **R**.

The ray parameter d can be approximated or found by an iterative process [147]. In the following subsections, we first discuss a simple approximation, then two iteration schemes that can be combined.

6.1.1 Parallax Correction

Classical environment mapping would look up the illumination selected by direction **R**, that is, it would use the radiance of point **r** (Figure 6.5). This can be considered as the first guess for the ray hit.

To find a better guess, we assume that the environment surface at **r** is perpendicular to the ray direction **R** [164, 166, 124]. In the case of a perpendicular surface, the ray would hit point **p**. Points **r**, **x**, and the origin **o** define a plane, which is the base plane of Figure 6.5. This plane also contains the visible point approximation **p** and the unit direction vector **R**. Multiplying ray equation

$$\mathbf{x} + \mathbf{R}d_p = \mathbf{p}$$

by the unit length direction vector **R** and substituting $\mathbf{R} \cdot \mathbf{p} = |\mathbf{r}|$, which is the consequence of the perpendicular surface assumption, we can express the ray parameter d_p as

$$d_p = |\mathbf{r}| - \mathbf{R} \cdot \mathbf{x}. \tag{6.1}$$

If we used the direction of point **p** to look up the environment map, we would obtain the radiance of point **p'**, which is in the direction of **p** but is on the surface.

The following HitParallax function gets the ray of origin x and direction R, as well as cube map distmap, which stores distance values in its alpha channel, and returns ray hit

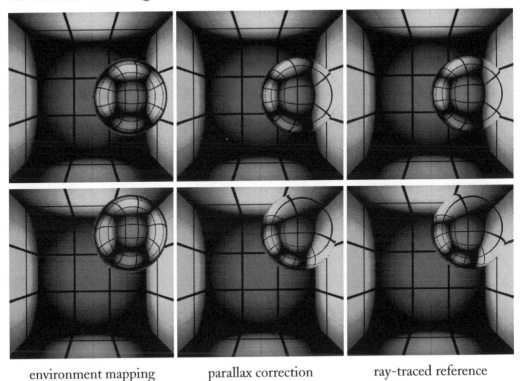

environment mapping   parallax correction   ray-traced reference

FIGURE 6.6: Comparison of classical environment mapping and parallax correction with ray-traced reflections placing the reference point at the center of the room and moving a reflective sphere close to the environment surface.

approximation p:

```
float3 HitParallax(in float3 x, in float3 R, sampler distmap) {
    float  rl = texCUBE(distmap, R).a; // |r|
    float  dp = rl - dot(x, R);
    float3 p = x + R * dp;
    return p;
}
```

6.1.2 Linear Search

The possible intersection points are on the half-line of the ray, thus intersection can be found by *marching* along the ray, i.e. checking points $\mathbf{l} = \mathbf{x} + \mathbf{R}d$ generated with an increasing sequence of parameter d, and detecting the first pair of subsequent points where one point is an overshooting while the other is an undershooting. The real intersection is between these two guesses, which is

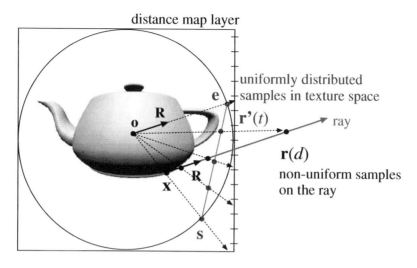

FIGURE 6.7: Cube map texels visited when marching along ray $\mathbf{x} + \mathbf{R}d$.

found by a secant search discussed in the following subsection. The definition of the increments of the ray parameter d needs special considerations since now the geometry is sampled, and it is not worth checking the same sample many times while ignoring other samples. Unfortunately, making uniform steps on the ray does not guarantee that the texture space is uniformly sampled. As we get farther from the reference point, constant length steps on the ray correspond to decreasing length steps in texture space. This problem can be solved by marching along a line segment that looks identical to the ray from the reference point, except that its two end points are at the same distance [147]. The end points of such a line segment can be obtained by projecting the start of the ray ($\mathbf{r}(0)$) and the end of the ray ($\mathbf{r}(\infty)$) onto a unit sphere, resulting in new start $\mathbf{s}$ and end point $\mathbf{e}$, respectively (Figure 6.7).

The intersection algorithm searches these texels, making uniform steps on the line segment of $\mathbf{s}$ and $\mathbf{e}$:

$$\mathbf{r}' = \mathbf{s}(1 - t) + \mathbf{e}t, \quad t = 0, \Delta t, 2\Delta t, \ldots, 1.$$

The correspondence between the ray parameter d and the line parameter t can be found by projecting $\mathbf{r}'$ onto the ray, which leads to the following formula:

$$d(t) = \frac{|\mathbf{x}|}{|\mathbf{R}|} \frac{t}{1 - t}.$$

The fragment shader implementation of the complete linear search gets the ray of origin x and direction R, as well as cube map distmap, and returns an undershooting ray parameter approximation dl and an overshooting approximation dp. Ratios $|\mathbf{l}|/|\mathbf{l'}|$ of the last undershooting and overshooting are represented by variables llp and ppp, respectively.

```
void HitLinear(in float3 x, in float3 R, sampler distmap, out float dl, out float dp) {
    float a = length(x) / length(R);
    bool  undershoot = false, overshoot = false;
    float llp;     // |l|/|l'| of last undershooting
    float ppp;     // |p|/|p'| of last overshooting

    float t = 0.0001f;  // Parameter of the line segment
    while( t < 1 && !(overshoot && undershoot) ) { // Iteration
        float d = a * t / (1 - t);  // Ray parameter corresponding to t
        float3 r = x + R * d;       // r(d): point on the ray
        float ra = texCUBElod(distmap, float4(r,0)).a; // |r'|
        float rrp = length(r)/ra;   // |r|/|r'|

        if (rrp < 1) {              // Undershooting
            dl = d;                 // Store last undershooting in dl
            llp = rrp;
            undershoot = true;
        } else {                    // Overshooting
            dp = d;                 // Store last overshooting as dp
            ppp = rrp;
            overshoot = true;
        }
        t += Dt;                    // Next texel
    }
}
```

This algorithm marches along the ray making uniform steps Dt in texture space. Step size Dt is set proportionally to the length of line segment of **s** and **e**, and also to the resolution of the cube map. By setting texel step Dt to be greater than the distance of two neighboring texels, we speed up the algorithm but also increase the probability of missing the reflection of a thin object. At a texel, the distance value is obtained from the alpha channel of the cube map. Note that we use function texCUBElod—setting the mipmap level explicitly to 0—instead of texCUBE, because texCUBE, like tex2D, would force the DirectX9 HLSL compiler to unroll the dynamic loop [119].

6.1.3 Refinement by Secant Search
Linear search provides the first undershooting and overshooting pair of subsequent samples, thus the real intersection must be between these points. Let us denote the overshooting and

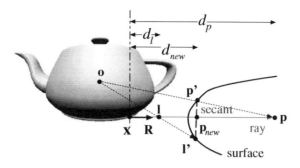

FIGURE 6.8: Refinement by a secant step.

undershooting ray parameters by d_p and d_l, respectively. The corresponding two points on the ray are $\mathbf{p}$ and $\mathbf{l}$, and the two points on the surface are $\mathbf{p}'$ and $\mathbf{l}'$, respectively (Figure 6.8).

Let us approximate the intersection of the surface and the plane of $\mathbf{o}, \mathbf{x}, \mathbf{R}$ by a line segment of points $\mathbf{l}'$ and $\mathbf{p}'$, defined by equation $\mathbf{l}'\alpha + \mathbf{p}'(1-\alpha)$. To find the intersection of this line and the ray, the following equation needs to be solved for the unknown ray parameter d_{new} and the line parameter α:

$$\mathbf{x} + \mathbf{R}d_{\text{new}} = \mathbf{l}'\alpha + \mathbf{p}'(1-\alpha).$$

Substituting identities

$$\mathbf{p}' = \mathbf{p}\frac{|\mathbf{p}'|}{|\mathbf{p}|} \quad \text{and} \quad \mathbf{l}' = \mathbf{l}\frac{|\mathbf{l}'|}{|\mathbf{l}|},$$

as well as the ray equation for d_p and d_l, we get

$$d_{\text{new}} = d_l + (d_p - d_l)\frac{1 - |\mathbf{l}|/|\mathbf{l}'|}{|\mathbf{p}|/|\mathbf{p}'| - |\mathbf{l}|/|\mathbf{l}'|}. \tag{6.2}$$

The point specified by this new ray parameter gets closer to the real intersection point. If a single secant step does not provide accurate enough results, then d_{new} can replace one of the previous approximations d_l or d_p, keeping always one overshooting approximation and one undershooting approximation, and we can proceed with the same iteration step.

The fragment shader implementation of the iterative secant refinement after an initial linear search is as follows:

```
float3 Hit(in float3 x, in float3 R, sampler distmap) {
    float  dl, dp;                     // Undershooting and overshooting ray parameters
    HitLinear(x, R, distmap, dl, dp);  // Find dl and dp with linear search
    for(int i = 0; i < NITER; i++) {   // Refine the solution with secant iteration
```

```
            dnew = dl + (dp-dl) * (1-llp)/(ppp-llp); // Ray parameter of new intersection
            float3 r = x + R * dnew;          // New point on the ray
            float rrp = length(r)/ texCUBElod(distmap, float4(r,0)).a; // |r|/|r'|
            if (rrp < 0.9999) {               // Undershooting
                llp = rrp;                    // Store as last undershooting
                dl = dnew;
            } else if (rrp > 1.0001) {        // Overshooting
                ppp = rrp;                    // Store as last overshooting
                dp = dnew;
            } else i = NITER;                 // Terminate the loop
        }
    return r;
}
```

Putting the linear search and the secant search together in a Hit function, we have a general tool to trace an arbitrary ray in the scene. This function finds the first hit l. Reading the cube maps in this direction, we can obtain the color of the hit point.

6.2   SINGLE LOCALIZED REFLECTIONS AND REFRACTIONS

In order to render objects specularly reflecting their environment, first we need to generate the distance map of the environment. The distance map stores the reflected radiance and the distance of points visible from the reference point at the directions of cube map texels. Having the distance map, we activate custom vertex and fragment shader programs and render the reflective objects. The vertex shader transforms the reflective object to clipping space, and also to the coordinate system of the cube map first applying the modeling transform, then translating the origin to the reference point. View vector **V** and normal **N** are also obtained in this space.

```
float3   refpoint;      // Reference point of the cube map
float3   EyePos;        // Eye position in world space

void SpecularReflectionVS(
    in  float4 Pos  : POSITION, // Reflection point in modeling space
    in  float3 Norm : NORMAL,   // Normal vector
    in  float2 Tex  : TEXCOORD0,// Texture coords
    out float4 hPos : POSITION, // Reflection point in clipping space
    out float3 x    : TEXCOORD1,// Reflection point in cube map space
    out float3 N    : TEXCOORD2,// Normal in cube map space
    out float3 V    : TEXCOORD3 // View in cube map space
) {
    hPos = mul(Pos, WorldViewProj);  // Transform to clipping space
    x    = mul(Pos, World) - refpoint;  // Transform to cube map space
    N    = mul(Norm, WorldIT);        // Transform normal to cube map space
    V    = x - EyePos;                // View in cube map space
}
```

Having the graphics hardware computed the homogeneous division and filled the triangle with linearly interpolating all vertex data, the fragment shader is called to find ray hit **l** and to look up the cube map in this direction.

The fragment shader calls function `Hit` to trace the ray and looks up cube map `distmap` again to find illumination `I` of the hit point. The next step is the computation of the reflection of incoming radiance `I`. If the surface is an ideal mirror, the incoming radiance should be multiplied by the Fresnel term evaluated for the angle between surface normal **N** and the reflection direction, or alternatively, the viewing direction. The Fresnel function is approximated according to equation (1.6).

```
samplerCUBE distmap; // distance map
float3      Fp;      // Fresnel at perpendicular illumination

float4 SpecularReflectionPS(
    float3 x : TEXCOORD1,    // Reflection point in cube map space
    float3 N : TEXCOORD2,    // Normal in cube map space
    float3 V : TEXCOORD3     // View in cube map space
) : COLOR
{
    V = normalize(V);
    N = normalize(N);
    float3 R = reflect(V, N);          // Reflection direction
    float3 l = Hit(x, R, distmap);     // Ray hit
    float3 I = texCUBE(distmap, l).rgb; // Incoming radiance
    // Fresnel reflection
    float3 F = Fp + pow(1-dot(N, -V), 5) * (1-Fp);
    return float4(F * I, 1);
}
```

Single refractions can be rendered similarly, but the direction computation should use the law of refraction instead of the law of reflection. In other words, the `reflect` operation should be replaced by the `refract` operation, and the incoming radiance should be multiplied by $1 - F$ instead of F. Figure 6.9 shows the results for a reflective/refractive glass skull and a reflective teapot put into a box. The knight of Figure 6.10 wears reflective armor rendered by the presented shaders.

During the application of the algorithm, the step size of the linear search and the iteration number of the secant search should be carefully selected since they can significantly influence the image quality and the rendering speed. Setting the step size of the linear search greater than the distance of two neighboring texels of the distance map, we can speed up the algorithm but also increase the probability that the reflection of a thin object is missed. If the geometry rendered into a layer is smooth, i.e. consists of larger polygonal faces, then the linear search can take larger steps and the secant search is able to find exact solutions taking just a few iterations.

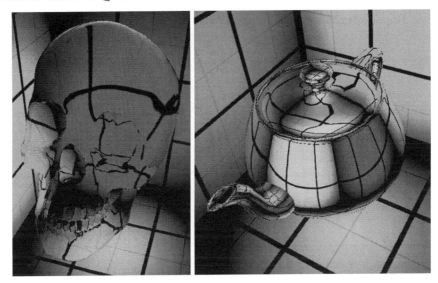

FIGURE 6.9: Left: a glass skull ($v = 1.3$) of 61 000 triangles rendered at 130 FPS. Right: an aluminum teapot of 2300 triangles rendered at 440 FPS.

FIGURE 6.10: A knight in reflective armor illuminated by dynamic lights.

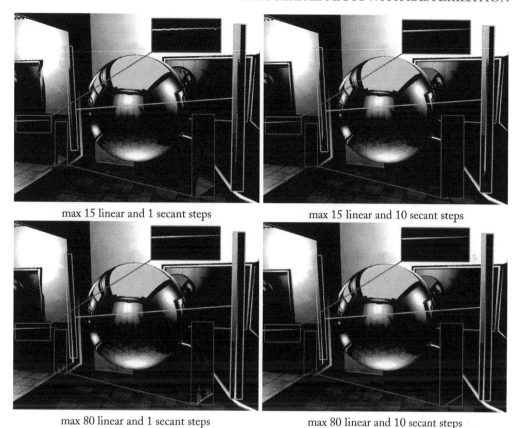

max 15 linear and 1 secant steps   max 15 linear and 10 secant steps

max 80 linear and 1 secant steps   max 80 linear and 10 secant steps

FIGURE 6.11: Aliasing artifacts when the numbers of linear/secant steps are maximized to 15/1, 15/10, 80/1, and 80/10, respectively.

However, when the distance value in a layer varies abruptly, the linear search should take fine steps in order not to miss spikes of the distance map, which correspond to thin reflected objects. Another source of undersampling artifacts is the limitation of the number of distance map layers and of the resolution of the distance map. In reflections we may see those parts of the scene which are coarsely or not at all represented in the distance maps due to occlusions and to their grazing angle orientation with respect to the center of the cube map. Figure 6.11 shows these artifacts and demonstrates how they can be reduced by appropriately setting the step size of the linear search and the iteration number of the secant search.

Note how the stair-stepping artifacts are generally eliminated by additional secant steps. Thin objects zoomed-in on the green and red frames require fine linear steps since if they are missed, the later secant search is not always able to quickly correct the error. Note that the secant search was more successful in the red frame than in the green frame since the table occupies more texels in the distance maps. The aliasing of the reflection of the shadowed area below the table in the left mirror, which is zoomed-in on the blue frame, is caused by the limitation of distance map layers. Since this area is not represented in the layers, not even the secant search can compute accurate reflections. Such problems can be addressed by increasing the number of distance map layers.

6.3   INTER-OBJECT REFLECTIONS AND REFRACTIONS

Cube map-based methods computing single reflections can straightforwardly be used to obtain multiple specular inter-reflections of different objects if each of them has its own cube map [99]. Suppose that the cube maps of specular objects are generated one by one. When the cube map of a particular object is generated, other objects are rendered into its cube map with their own shader. A diffuse object is rendered with the reflected radiance of the direct light sources, and a specular object is processed by the discussed fragment shader that looks up its cube map in the direction of the hit point of the reflection (or refraction) ray. When the first object is processed the cube maps of other objects are not yet initialized, so the cube map of the first object will be valid only where diffuse surfaces are visible. However, during the generation of the cube map for the second reflective object, the color reflected off the first object is already available, thus *diffuse surface–first reflective object–second reflective object* kind of paths are correctly generated. At the end of the first round of the cube map generation process, a later generated cube map will contain the reflection of other reflectors processed earlier, but not vice versa. Repeating the cube map generation process again, all cube maps will store double reflections and later rendered cube maps also represent triple reflections of earlier processed objects. Cube map generation cycles should be repeated until the required reflection depth is reached. If we have a dynamic scene when cube maps are periodically re-generated anyway, the calculation of higher order reflections is for free (Figure 6.12). The reflection of a reflected image might come from the previous frame, but this delay is not noticeable at interactive frame rates.

Note that this approach is not able to simulate intra-object or *self reflections*, neither can it solve inter-reflections for a group of objects that are assigned to the same cube map. Such self reflections require recursive ray-tracing, and are discussed in Section 6.6.

6.4   SPECULAR REFLECTIONS WITH SEARCHING ON THE REFLECTOR

The discussed algorithm used visibility ray-tracing, and having identified the point visible from the camera, i.e. the reflection point, it started to look for the reflected point in the distance map. Searching the reflected environment stored in cube maps is not the only way to render reflections. For example, we can start at the reflected points, i.e. at the vertices of

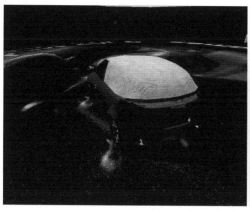

FIGURE 6.12: Inter-object reflections in RT car game. Note the reflection of the reflective car on the beer bottles.

the environment, and search for the reflection points, i.e. identify those points that reflect the input points toward the camera. This means searching on the reflector surface rather than on the environment surface [38, 118, 37].

The comparative advantages of searching on the environment surface or searching on the reflector depend on the geometric properties of these surfaces. The algorithm searching on the environment surface can handle arbitrary reflector surfaces and is particularly efficient if the environment surface does not have large distance variations. On the other hand, the algorithm searching on the reflector surface can cope with environment surfaces of large depth variations, but is limited to simple reflectors that are either concave or convex.

6.5 SPECULAR REFLECTIONS WITH GEOMETRY OR IMAGE TRANSFORMATION

Planar mirrors used to be rendered by mirroring virtual objects through the plane for each reflection and blending their picture into the image [33]. Pre-computed radiance maps [13] can be used to speed up the process. For curved reflectors, the situation is more complicated. The method of Ofek and Rappoport [100] warps the surrounding scene geometry such that it appears as a correct virtual image when drawn over the top of the reflector. For each vertex to be reflected, an efficient data structure (the explosion map) accelerates the search for a triangle which is used to perform the reflection. An analytic approach has also been developed [22], using a preprocessing step based on the path perturbation theory.

Making images dependent on the viewer's position is also important in *image-based rendering* [88]. Image-based rendering can be explained as combining and warping images taken from different camera positions in order to get the image for a new camera position. The reconstruction may be based on coarse geometric information [48, 31] or on per-pixel depth values [108, 122, 40]. Lischinski and Rappoport [90] used layered light fields to render fuzzy reflections on glossy objects and layered depth images were used to ray-trace sharp reflections. Heidrich et al. [64] and Yu et al. [167] used two light fields to simulate accurate reflections, one for the radiance of the environment, and another for mapping viewing rays striking the object to outgoing reflection rays.

6.6   SELF REFLECTIONS AND REFRACTIONS

Light may get reflected or refracted on an ideal reflector or refractor several times. If the hit point of a ray is again on the specular surface, then reflected or refracted rays need to be computed and ray-tracing should be continued, repeating the same algorithm recursively. The computation of the reflected or refracted ray requires the normal vector at the hit surface, the Fresnel function, and also the index of refraction in the case of refractions. These attributes can also be stored in distance maps.

If *self reflections* and refractions need to be computed, we cannot decompose the scene to the reflector and to its environment, but the complete geometry must be stored in distance maps. On the other hand, we cannot ignore those points that are occluded from the reference point, which means that a distance map texel should represent a set of points that are at the same direction from the reference point. We may store a list of distances in a distance map texel or several distance maps may be used, where each of them represents a single depth *layer*. Assigning objects to layers is a non-trivial process.

The set of layers representing the scene could be obtained by *depth peeling* [141, 39, 89] in the general case (see Section 10.4 for an implementation of the depth peeling algorithm). However, in many cases a much simpler strategy may also work.

If we simulate only multiple refractions on a single object, then the reference point can be put to the center of the refractor, and a single cube map layer can be assigned to the refractor surface, and one other for the environment, thus two layers can solve the occlusion problem (Figure 6.13). On the other hand, in the case of a reflecting surface, we may separate the specular surface from its environment, place the reference point close to the specular object, and assign one layer for the front faces of the specular surface, one for the back faces, and a third layer for the environment.

We use two cube maps for each layer. The first cube map includes the material properties, i.e. the reflected color for diffuse surfaces or the Fresnel function at perpendicular illumination for specular surfaces, and the index of refraction. The second cube map stores the normal vectors and the distance values. To distinguish ideal reflectors, refractors, and diffuse surfaces, the sign

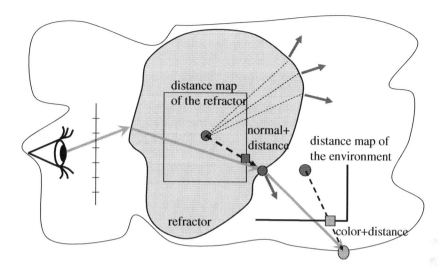

FIGURE 6.13: Computation of multiple refractions on a single object storing the object's normals in one distance map and the color of the environment in another distance map.

of the index of refraction is checked. Negative, zero, and positive values indicate a reflector, a diffuse surface, and a refractor, respectively.

Generally, searching multi-layer distance maps is equivalent to searching every layer with the discussed single-layer algorithms and then selecting the closest intersection. However, in special cases, we can reduce the computational burden of multiple layers. Supposing that objects do not intersect each other, a ray refracted into the interior of an object must meet its surface again before hitting the environment surface, thus it is enough to search only the layer of the refractor surface (Figures 6.13 and 6.14). However, when the images of Figure 6.15 were rendered we needed to maintain three layers.

In the implementation we shall assume that the `Hit` function discussed so far is modified, renamed to `HitMultiLayer`, and now it is able to search in multiple layers and find the closest intersection. With this function, the fragment shader computing multiple specular effects organizes the ray-tracing process in a dynamic loop. The fragment shader should find the ray hit and look up the cube map in this direction. It calls function `HitMultiLayer` that implements the combination of linear and secant searches and finds the first hit `l`, its radiance `Il`, and normal vector `Nl`. If the surface is an ideal mirror, the incoming radiance should be multiplied by the Fresnel term evaluated for the angle between the surface normal and the viewing direction.

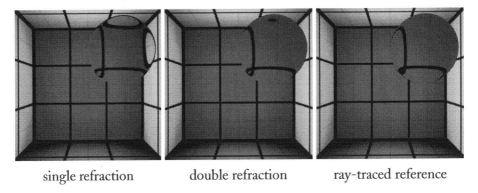

single refraction   double refraction   ray-traced reference

FIGURE 6.14: Single and multiple refractions on a sphere having refraction index $\nu = 1.1$.

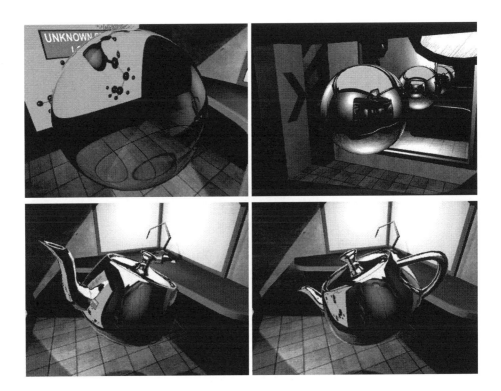

FIGURE 6.15: Multiple refractions and reflections when the maximum ray depth is four.

```
float4 MultipleReflectionPS(
    float3 x : TEXCOORD1,    // Reflection point in cube map space
    float3 N : TEXCOORD2,    // Normal vector
    float3 V : TEXCOORD3,    // View direction
    uniform float3 Fp0,      // Fresnel at perpendicular direction
    uniform float3 n0        // Index of refraction
) : COLOR
{
    V = normalize(V); N = normalize(N);
    float3 I = float3(1, 1, 1);// Radiance of the path
    float3 Fp = Fp0;           // Fresnel at 90 degrees at first hit
    float  n = n0;             // Index of refraction of the first hit
    int depth = 0;             // Length of the path
    while (depth < MAXDEPTH) {
        float3 R;         // Reflection or refraction dir
        float3 F = Fp + pow(1-abs(dot(N, -V)), 5) * (1-Fp);
        if (n <= 0) {
            R = reflect(V, N);  // Reflection
            I *= F;             // Fresnel reflection
        } else {                // Refraction
            if (dot(V, N) > 0) {// Coming from inside
                n = 1 / n;
                N = -N;
            }
            R = refract(V, N, 1/n);
            if (dot(R, R) == 0) // No refraction direction exits
                    R = reflect(V, N); // Total reflection
            else    I *= (1-F); // Fresnel refraction
        }
        float3 Nl;              // Normal vector at the hit point
        float4 Il;              // Radiance at the hit point
        // Trace ray x+R*d and obtain hit l, radiance Il, normal Nl
        float3 l = HitMultiLayer(x, R, Il, Nl);
        n = Il.a;
        if (n != 0) {       // Hit point is on specular surface
            Fp = Il.rgb;    // Fresnel at 90 degrees
            depth += 1;
        } else {            // Hit point is on diffuse surface
            I *= Il.rgb;    // Multiply with the radiance
            depth = MAXDEPTH; // Terminate
        }
        N = Nl; V = R; x = l; // Hit point is the next shaded point
    }
    return float4(I, 1);
}
```

6.6.1 Simplified Methods for Multiple Reflections and Refractions

Wyman [163] proposed a front face/back face double refraction algorithm. During pre-computation this method calculates the distance of the back of the refractor object at the direction of the normal vector at each vertex. In the first pass of the on-line part, normals and depths of back-facing surfaces are rendered into a texture. In the second pass front faces are drawn. The fragment shader reads the pre-computed distance in the direction of the normal and obtains the distance in the view direction from the texture, and linearly interpolates these distances according to the angle between the view and normal directions and the angle between the refracted and the normal directions. Translating the processed point by the distance obtained by this interpolation into the refraction direction, we get an approximation of the location of the second refraction. Projecting this point into the texture of rendered normal vectors, the normal vector of the second refraction is fetched and is used to compute the refracted direction of the second refraction.

6.7 CAUSTICS

Caustics show up as beautiful patterns on diffuse surfaces, formed by light paths originating at light sources and visiting mirrors or refracting surfaces first (Figure 6.16). A caustic is the concentration of light, which can "burn". The name caustic comes from the Latin "causticus" derived from Greek "kaustikos", which means "burning".

Caustic rendering should simulate *specular light paths* starting at the light sources and visiting specular reflectors and refractors until they arrive at a diffuse surface. Caustic generation

FIGURE 6.16: Real caustics.

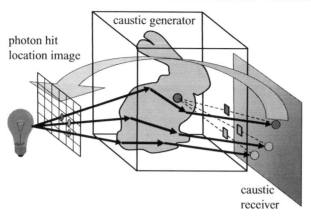

FIGURE 6.17: The light pass renders into the photon hit location image where each pixel stores the location of the photon hit.

algorithms have more than one phase [5]. First, the scene is processed from the point of view of the light source, and the terminal hits of specular paths are determined. Secondly, photon hits are splat and blurred to produce continuous caustic patterns, which are projected onto a *light map*[1], texture map, or directly onto the image. If the caustic patterns are generated in a light map or texture map, then the next camera pass will map them onto the surfaces and present them to the user.

6.7.1 Light Pass

The first pass of the caustic generation is the identification of terminal hits of specular light paths originating at the light source. In this pass, the reflective or refractive objects, called *caustic generators*, are rendered from the point of view of the light source. In this phase, the view plane is placed between the light and the refractor (Figure 6.17). The pixels of the window of this rendering pass correspond to a dense sampling of light rays emitted by the source. If this rendering pass is implemented on the GPU, then the fragment shader should generate the specular light path, implementing some form of ray-tracing. We can use the algorithms of GPU-based ray-tracing (Chapter 5) or the distance map-based approximate strategy working on the sampled geometry (Section 6.1). For a single pixel, the results of this pass are the final hit point and the power of the specular light path going through this pixel, which are written

[1]A light map is a texture map that stores irradiance values. To get the visible colors, the irradiance of a texel should be multiplied by the BRDF usually stored in another texture map.

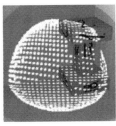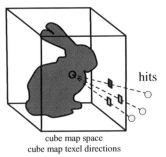

screen space   texture space   cube map space
pixel coordinates   texel coordinates   cube map texel directions

FIGURE 6.18: The three alternatives of storing photon hit locations. Screen space methods store pixel coordinates, texture space methods texture coordinates. Cube map space methods store the direction of the photon hit with respect to the center of a cube map associated with the caustic generator (the bunny in the figure).

into the R, G, B, A channels of the render target. The resulting image may be called the *photon hit location image*.

Since discrete photon hits should finally be combined to a continuous caustic pattern, a photon hit should affect not only a surface point, but also a surface neighborhood where the power of the photon is distributed. In order to eliminate light leaks, the neighborhood also depends on whether its points are visible from the caustic generator object. Finally, the computation of the reflection of the caustics illumination requires the local BRDF, thus we should find an easy way to reference the surface BRDF with the location of the hit. There are several alternatives to represent the location of a photon hit (Figure 6.18):

Texture space [137]: Considering that the reflected radiance caused by a photon hit is the product of the BRDF and the power of the photon, the representation of the photon hit should identify the surface point and its BRDF. A natural identification is the texture coordinates of that surface point which is hit by the ray. A pixel of the photon hit location image stores two texture coordinates of the hit position and the luminance of the power of the photon. The photon power is computed from the power of the light source and the solid angle subtended by the caustic map pixel. Since the texture space neighborhood of a point visible from the caustic generator may also include occluded points, light leaks might occur.

Screen or image space [166]: A point in the three-dimensional space can be identified by the pixel coordinates and the depth when rendered from the point of view of the camera. This screen space location can also be written into the photon hit location image. If photon hits are represented in image space, photons can be splat directly onto the image of the diffuse caustic receivers without additional transformations. However, the BRDF of the surface point cannot be easily looked up with this representation, and we should modulate the rendered color with the caustic light, which is only an approximation. This method is also prone to creating light leaks.

Cube map space [124]: The coordinate system of the distance map used to trace rays leaving the caustic generator automatically provides an appropriate representation. A point is identified by the direction in which it is visible from the reference point, i.e. the texel of the cube map, and also by the distance from the reference point. An appropriate neighborhood for filtering is defined by those points that are projected onto neighboring texels taking the reference point as the center of projection, and having similar distances from the reference point as stored in the distance map. Note that this approach is very similar to classical shadow mapping and successfully eliminates light leaks.

In order to recognize those texels of the photon hit location image where the refractor is not visible, we initialize this image with −1 alpha values. Checking the sign of the alpha later, we can decide whether or not it is a valid photon hit.

6.7.2   Photon Hit Filtering and Light Projection

Having the photon hit location image, blurred photon hits should modify the irradiance of the receiver surfaces. Since we usually work with fewer photons than what would create a continuous pattern, this process also involves some filtering or blurring. We note that filtering can be saved if we take high number of photons and find their hit point with a pixel precision method [84].

Most of the GPU-based caustics algorithms [153, 137, 124, 166, 165] blurred photon hits by photon *splatting*. On the other hand, we can also use *caustic triangles* instead of splatting, where the vertices of a triangle are defined by three neighboring photon hits.

Now we discuss photon splatting in details. During this pass photon hits are rendered, taking their actual position from the photon hit location image. We send as many small quadrilaterals (two adjacent triangles in Direct3D) or point sprites as the photon hit location image has (Figure 6.19) down the pipeline. We can avoid the expensive GPU to CPU transfer of this image, if the vertex shader modulates the location of points by the content of the photon hit location image. This operation requires at least Shader Model 3 GPUs that allow the vertex shader to access textures.

Rectangles with the corrected position are rasterized, and its fragments are processed by the fragment shader. In order to splat the photon on the render target, a splatting filter texture is associated with the rectangle and alpha blending is turned on to compute the total contribution of different photon hits.

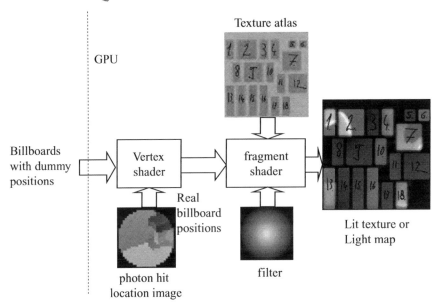

FIGURE 6.19: Photon hit filtering pass assuming that the photon hit location image stores texture coordinates.

Let us now consider the case when photon hits are represented in texture space and the render target of this operation is the texture atlas or a light map. Rectangles arrive at the vertex shader with dummy position and a serial number `caustCoord` that specifies that photon hit which will provide the location of this rectangle (Figure 6.19). The photon hit location texture is addressed with `caustCoord` in the vertex shader shown below. The size of the rectangle depends on the width of the splatting filter. The vertex shader changes the coordinates of the quadrilateral vertices and centers the quadrilateral at the u, v texture coordinates of the photon hit in texture space if the alpha value of the caustic map texel addressed by `caustCoord` is positive, and moves the quadrilateral out of the clipping region if the alpha is negative (i.e. it is not a valid photon hit).

The vertex shader of projecting caustic patterns onto the texture atlas is as follows:

```
void CausticRenderVS(
    in  float4 Pos        : POSITION,    // Vertex of a small quad
    in  float2 caustCoord : TEXCOORD0,   // Address for the photon hit location image
    out float4 hPos       : POSITION,    // Photon hit transformed to clipping space
    out float2 filtCoord  : TEXCOORD1,   // Texture coord used by filtering
    out float  Power      : TEXCOORD2,   // Incident power of the hit
    out float4 Tex        : TEXCOORD3)   // Texture coords to address BRDF data
{
```

```
    // Photon position fetched from the photon hit location image
    float4 ph = tex2Dlod(caustmap, caustCoord);
    filtCoord = Pos.xy; // Filter coords
    // Place quad vertices in texture space to fetch BRDF data
    Tex.x = ph.x + Pos.x / 2;
    Tex.y = ph.y - Pos.y / 2;
    // Transform photon rectangle to clipping space which is identical to texture space
    hPos.x = ph.x * 2 - 1 + Pos.x + HALF;
    hPos.y = 1 - ph.y * 2 + Pos.y - HALF;
    if (ph.a > 0 ) hPos.z = 0; // It is a real hit
    else           hPos.z = 2; // It is not a real hit -> Ignore
    hPos.w = 1;
    Power = ph.a; // Pass incident power
}
```

Note that the original x, y coordinates of quadrilateral vertices are copied as filter texture coordinates, and are also moved to the position of the photon hit in the texture space of the surface. The output position register (hPos) also stores the texture coordinates converted from $[0, 1]^2$ to $[-1, 1]^2$ which corresponds to rendering to this space. The w and z coordinates of the position register are used to ignore those photon hit location texture elements which have no associated valid photon hit.

The fragment shader computes the color contribution as the product of the photon power, filter value, and the BRDF:

```
float4 CausticRenderPS(
    float2 filtCoord : TEXCOORD1,    // Texture coords for Gaussian filter
    float  Power     : TEXCOORD2,    // Incident power of the hit
    float4 Tex       : TEXCOORD3 ) : COLOR
{
    float4 brdf = tex2D(brdfmap, Tex);
    float4 w = tex2D(filter, filtCoord );
    return power * w * brdf;
}
```

The target of this pass is the light map or the modified texture map. Note that the contributions of different photons should be added, thus we should set the blending mode to "add" before executing this phase (Figure 6.20). When the scene is rendered from the point of view of the camera but with the modified textures, caustic patterns will show up in the image. This means that the conventional texturing process will project the caustic patterns onto the surfaces.

FIGURE 6.20: A photon hit location image, a room rendered without blending, and the same scene with blending enabled.

If photon hits are represented in image space, then the render target is the frame buffer, and caustic patterns are directly placed over a rendered image, so no separate light projection phase is necessary.

If photon hits are defined in cube map space, then blurring is actually realized on cube map faces resulting in a *light cube map* (Figure 6.21). During the camera pass, we can project these patterns onto the surfaces similarly to shadow mapping algorithms [124]. The processed point is transformed into the coordinate system of the cube map and the direction

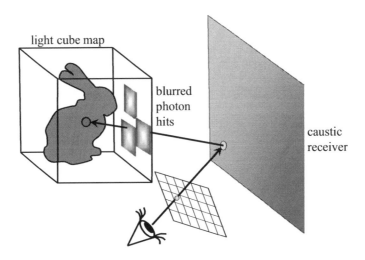

FIGURE 6.21: Light projections pass assuming the the photon hit location image stores cube map texel directions.

FIGURE 6.22: Caustics seen through the refractor object.

from the cube map center to the processed point is obtained. The cube map texels store distance values. The distance associated with the direction is fetched and is compared to the distance of the processed point. If these values are similar, then the processed point is visible from the point of view of the caustic generator, so no object prohibits the processed point to receive the extra illumination of the caustics, that is stored in the cube map texel. Although this approach requires one additional texture lookup in the shader, it eliminates light leaks caused by splatting.

FIGURE 6.23: Real-time caustics caused by glass objects (index of refraction $\nu = 1.3$).

FIGURE 6.24: Reflective and refractive spheres in a game environment.

6.8 COMBINING DIFFERENT SPECULAR EFFECTS

The different techniques generating reflection, refraction, and caustics can be combined in a complete real-time rendering algorithm. The input of this algorithm includes

1. the definition of the diffuse environment in form of triangle meshes, material data, textures, and light maps having been obtained with a global illumination algorithm,

2. the definition of specular objects set in the actual frame,

3. the current position of light sources and the eye.

FIGURE 6.25: Reflective and refractive spheres in PentaG (http://www.gebauz.com) and in Jungle-Rumble (http://www.gametools.org/html/jungle_rumble.html).

FIGURE 6.26: The Space Station game rendered with the discussed reflection, refraction, and caustic method in addition to computing diffuse interreflections (left), and compared to the result of the local illumination model (right).

The image generation consists of a preparation phase, and a rendering phase from the eye. The preparation phase computes the distance maps. Depending on the distribution of specular objects, we may generate only a single cube map for all of them, or we may maintain a separate map for each of them. Note that this preparation phase is usually not executed in each frame, only if the object movements are large enough. If we update these maps after every 100 frames, then the cost amortizes and the slowdown becomes negligible. If the scene has caustic generators, then a photon hit location image is generated for each of them, which are converted to light maps during the preparation phase.

The final rendering phase from the eye position consists of three steps. First the diffuse environment is rendered with their light maps making also caustics visible. Then specular objects are sent to the GPU, having enabled the discussed reflection and refraction shaders. Note that in this way the reflection or refraction of caustics can also be generated (Figures 6.22 and 6.23).

The discussed methods have been integrated in games (Figures 6.24, 6.25, and 6.26, and refer also to Chapter 12 and Figures 12.3, 12.8, and 12.13). In these games $6 \times 256 \times 256$ resolution cube maps were used. It was also noticed that the speed improves by 20% if the distance values are separated from the color data and stored in another texture map. The reason of this behavior is that the pixel shader reads the distance values several times from different texels before the color value is fetched, and separating the distance values increases texture cache utilization.

CHAPTER 7

Diffuse and Glossy Indirect Illumination

Optically rough surfaces reflect light not just at a single direction as mirror-like smooth surfaces but at infinitely many directions. The reflected radiance of very rough surfaces, which are called *diffuse* surfaces, is similar for all viewing directions. Smoother, *glossy* surfaces still reflect in all possible directions but the radiance is higher in the vicinity of the ideal reflection direction. At a diffuse or glossy surface, a light path may be continued at infinitely many directions (Figure 7.1), from which a high enough number of sample paths should be computed and their contributions should be added.

GPU native local illumination algorithms compute a single light bounce. Since global illumination requires multiple bounces, the results of a single bounce should be stored and the whole computation is repeated. The main problem is the temporary storage of the illumination information, which requires data structures to be stored on the GPU.

A *color texture* of the surface can be interpreted as the radiance function sampled at points corresponding to the texel centers [12, 98]. An elementary surface area $A^{(i)}$ is the surface which is mapped onto texel i. The radiance is valid in all outgoing directions for diffuse surfaces, or can be considered as sampled at a single direction per each point if the surfaces are non-diffuse [11]. Alternatively, the texture can be viewed as the finite-element representation of the radiance function using piecewise constant or linear basis functions depending on whether bi-linear

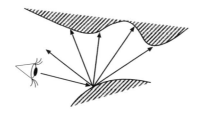

FIGURE 7.1: At diffuse or glossy surfaces light paths split.

texture filtering is enabled or not. Such radiance representation is ideal for obtaining the image from a particular camera since the texturing algorithm of the graphics hardware automatically executes this task.

In this chapter, we examine three techniques in detail, a GPU-based on-line stochastic radiosity algorithm [11], which uses color texture radiance representation, the application of pre-computed radiosity solutions, and a cube map-based final gathering method [139], which also stores distance information in texels, and allows the application of the stored illumination information more accurately than diffuse/glossy environment mapping.

7.1   RADIOSITY ON THE GPU

Global illumination methods considering only diffuse surfaces and lights are often called *radiosity* methods. The iterative *radiosity* method is one of the oldest global illumination techniques, and is also a pioneer of the exploitation of the graphics hardware [25]. It has been recognized for a long time that the *form factors*[1] belonging to a single point can be obtained by rendering the scene from this point, for which the graphics hardware is an ideal tool. In the era of non-programable GPUs the result of the rendering, i.e. the image, was read back to the CPU where the form factor computation and the radiosity transfer took place. The programmability of GPUs, however, offered new possibilities for radiosity as well. Now it is feasible to execute the complete algorithm on the GPU, without expensive image readbacks.

Let us revisit the reflected radiance formula (equation (1.8)) for diffuse surfaces. Now we express it as an integral over the surfaces, and not over the illumination directions:

$$L^r(\mathbf{x}) = \mathcal{T}_{f_r} L = \int_S L(\mathbf{y}) f_r(\mathbf{x}) G(\mathbf{x}, \mathbf{y}) \, d\mathbf{y}. \tag{7.1}$$

Here, S is the set of surface points, $f_r(\mathbf{x})$ is the diffuse BRDF of point $\mathbf{x}$, and

$$G(\mathbf{x}, \mathbf{y}) = v(\mathbf{x}, \mathbf{y}) \frac{\cos^+ \theta_{\mathbf{x}} \cos^+ \theta_{\mathbf{y}}}{|\mathbf{x} - \mathbf{y}|^2}$$

is the *geometric factor*, where $v(\mathbf{x}, \mathbf{y})$ is the mutual visibility indicator which is 1 if points $\mathbf{x}$ and $\mathbf{y}$ are visible from each other and zero otherwise, $\theta_{\mathbf{x}}$ and $\theta_{\mathbf{y}}$ are the angles between the surface normals and direction $\omega_{\mathbf{y} \to \mathbf{x}}$ that is between $\mathbf{x}$ and $\mathbf{y}$.

If the emission radiance is L^e, the first bounce illumination is $\mathcal{T}_{f_r} L^e$, the second bounce illumination is $\mathcal{T}_{f_r} \mathcal{T}_{f_r} L^e$, etc. The full global illumination solution is the sum of all bounces.

---

[1]The *point to polygon form factor* is the fraction of the power emitted by a differential surface at the point that arrives at the polygon.

Iteration techniques take some initial reflected radiance function, e.g. $L_0^r = 0$, and keep refining it by computing the reflection of the sum of the emission and the previous reflected radiance estimate. Formally, the global illumination solution which includes paths of all lengths is the limiting value of the following iteration scheme:

$$L_m^r = T_{f_r} L_{m-1} = T_{f_r}(L^e + L_{m-1}^r),$$

where L_m^r is the reflected radiance after iteration step m.

Each iteration step adds the next bounce to the reflected radiance. Iteration works with the complete radiance function, whose temporary version should be represented somehow. The classical approach is the *finite-element method*, which approximates the radiance function in a function series form. In the simplest diffuse case we decompose the surface to small elementary surface patches $A^{(1)}, \ldots, A^{(n)}$ and apply a piecewise constant approximation, thus the reflected radiance function is represented by the average reflected radiance of these patches, that is by $L^{r,(1)}, \ldots, L^{r,(n)}$ (here $L^{r,(i)}$ is the reflected radiance of patch i). CPU radiosity algorithms usually decompose surfaces to triangular patches. However, in GPU approaches this is not feasible since the GPU processes patches independently thus the computation of the interdependence of patch data is difficult. Instead, the radiance function can be stored in a texture, thus the elementary surfaces will correspond to different texels.

If the elementary surfaces are small, we can consider just a single point of them in the algorithms, while assuming that the properties of other points in the patch are similar. Surface properties, such as the BRDF and the emission can be given by values $f_r^{(1)}, \ldots, f_r^{(n)}$, and $L^{e,(1)}, \ldots, L^{e,(n)}$, respectively, in each texel. In the case of small elementary surfaces, we can check the mutual visibility of two elementary surfaces by inspecting only their centers. In this case, the update of the *average* reflected radiance corresponding to texel i in a single iteration can be approximated in the following way:

$$L_m^{r,(i)} = \frac{1}{A^{(i)}} \int_{A^{(i)}} L_m^r(\mathbf{x}) \, dx = \frac{1}{A^{(i)}} \int_{A^{(i)}} \int_S (L^e(\mathbf{y}) + L_{m-1}^r(\mathbf{y})) f_r^{(i)} G(\mathbf{x}, \mathbf{y}) \, dy \, dx \approx$$

$$\sum_{j=1}^n (L^{e,(j)} + L_{m-1}^{r,(j)}) f_r^{(i)} G(\mathbf{x}_i, \mathbf{y}_j) A^{(j)}, \qquad (7.2)$$

where $\mathbf{y}_j$ and $\mathbf{x}_i$ are the centers of elementary surfaces $A^{(j)}$ and $A^{(i)}$ belonging to texels j and i, respectively.

Iteration simultaneously computes the interaction between all surface elements, which has quadratic complexity in terms of the number of finite elements, and its GPU implementation turned out not to be superior than the CPU version [20]. The complexity of a single iteration step can be attacked by special iteration techniques, such as Southwell iteration (also called *progressive radiosity*) [152], hierarchical radiosity [54, 9, 14], or by randomization [125, 94]. Southwell iteration computes the interaction of the element having the highest unshot power and all other surface elements [24]. It is quite simple to implement on the GPU [26], but has also quadratic complexity [140], which limits these approaches for simple models and low texture resolutions.

Monte Carlo techniques, on the other hand, have sub-quadratic complexity, and can greatly benefit from the hardware. The method discussed here is based on the *stochastic iteration* [135, 94, 120, 14]. Stochastic iteration means that in the iteration scheme a *random transport operator* $T_{f_r}^*$ is used instead of the light-transport operator T_{f_r}. The random transport operator has to give back the light-transport operator in the expected case:

$$L_m^r = T_{f_r}^*(L^e + L_{m-1}^r), \qquad E[T_{f_r}^* L] = T_{f_r} L.$$

Note that such an iteration scheme does not converge, but the iterated values will fluctuate around the real solution. To make the sequence converge, we compute the final result as the average of the estimates of subsequent iteration steps:

$$L^r = \frac{1}{m} \sum_{k=1}^{m} L_k^r.$$

The core of all stochastic iteration algorithms is the definition of the random transport operator. We prefer those random operators that can be efficiently computed on the GPU and introduce small variance. Note that randomization gives us a lot of freedom to define an elementary step of the iteration. This is very important for GPU implementation since we can use iteration steps that fit to the features of the GPU.

One approach meeting these requirements is the *random hemicube shooting* (*perspective ray bundles*) [136], which selects a patch (i.e. a texel) randomly, and its radiance is shot toward other surfaces visible from here (Figure 7.2). Note that this elementary step is similar to that of progressive radiosity. However, while progressive radiosity selects that patch which has the highest unshot power, and shoots this unshot power, stochastic hemicube shooting finds a patch randomly, and shoots the current estimate.

Note that random hemicube shooting is equivalent to the Monte Carlo evaluation of the sum of equation (7.2). A surface element $A^{(j)}$ is selected with probability p_j, and y_j is set to its center. This randomization allows us to compute the interaction between shooter surface element $A^{(j)}$ and all other receiver surface elements $A^{(i)}$, instead of considering all shooters and receivers simultaneously. Having selected shooter $A^{(j)}$ and its center point y_j, the radiance of this point is sent to all those surface elements that are visible from here. The Monte Carlo estimate of the reflected radiance of surface element i after this transfer is the ratio of the

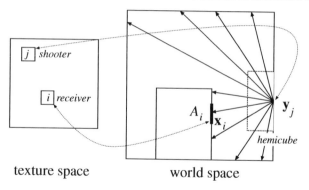

texture space   world space

FIGURE 7.2: A single iteration step using random hemicube shooting from $\mathbf{y}_j$.

transferred radiance and the probability of this sample p_j, that is

$$\tilde{L}_m^{r,(i)} = \frac{\left(L^{e,(j)} + L_{m-1}^{r,(j)}\right) f_r^{(i)} G(\mathbf{x}_i, \mathbf{y}_j) A^{(j)}}{p_j}. \qquad (7.3)$$

In order to realize this random transport operator, two tasks need to be solved. The first is the random selection of a texel identifying point $\mathbf{y}_j$. The second is the update of the radiance at those texels which correspond to points $\mathbf{x}_i$ visible from $\mathbf{y}_j$.

7.1.1   Random Texel Selection

The randomization of the iteration introduces some variance in each step, which should be minimized in order to get an accurate result quickly. According to *importance sampling*, the introduced variance can be reduced with a selection probability that is proportional to the integrand or to the sampled term of the sum of equation (7.2). Unfortunately, this is just approximately possible, and the selection probability is set proportional to the current power of the selected texel. The power is proportional to the product of the area and the radiance of the corresponding surface patch, i.e. to $(L^{e,(j)} + L_{m-1}^{r,(j)}) A^{(j)}$. If the light is transferred on several wavelengths simultaneously, the *luminance* of the power should be used.

Let us imagine the luminance of a patch as the length of an interval and assume that these intervals are put after each other forming a large interval of length equal to the total luminance of the scene. Selecting a patch proportionally to its luminance is equivalent to generating a uniformly distributed random number r in the large interval and returning that patch whose interval contains the random point.

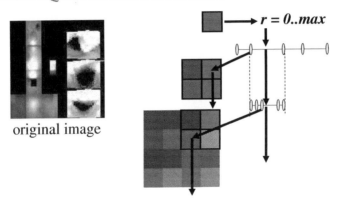

FIGURE 7.3: Random texel selection using mipmapping.

The random selection proportional to the luminance can also be supported by a *mipmapping* scheme. A mipmap can be imagined as a quadtree, which allows the selection of a texel in $\log_2 \mathcal{R}$ steps, where $\mathcal{R}$ is the resolution of the texture. Each texel is the sum (or the average) of the luminance of four corresponding texels on a lower level. The top level of this hierarchy has only one texel, which contains the sum of the luminance of all elementary texels (Figure 7.3). Both the generation and the sampling require the rendering of a textured rectangle that covers all pixels (also called a viewport sized or full-screen quad) by $\log_2 R$ times.

The generated mipmap is used to sample a texel with a probability that is proportional to its luminance. First the total luminance of the top-level texel is retrieved from a texture and is multiplied by a random number uniformly distributed in the unit interval. Then the next mipmap level is retrieved, and the four texels corresponding to the upper level texel are obtained. The coordinates of the bottom right texel are passed in variable uv, and the coordinates of the three other texels are computed by adding parameter rr that is the distance between two neighboring texels. The luminance of the four texels are summed until the running sum (denoted by cmax in the program below) gets greater than selection value r obtained on the higher level. When the running sum gets larger than the selection value from the higher level, the summing is stopped and the actual texel is selected. A new selection value is obtained as the difference of the previous value and the luminance of all texels before the found texel (r-cmin). The new selection value and the texture coordinates of the selected texel are returned in c. Then the same procedure is repeated in the next pass on the lower mipmap levels. This procedure terminates at a leaf texel with a probability that is proportional to its luminance. The shader of a pass of the mipmap-based sampling may call the following function to select a texel according to random selection value r:

```
void SelectTexel(inout float   r,   // Random selection value
                 inout float2 uv,   // Right-bottom texel of the four texel group
                 in    float   rr)  // Two neighboring texels in texture space
{
    float cmin = 0;                             // Running sum so far
    float cmax = tex2D(texture, uv);            // New running sum
    if(cmax >= r) {
        r = r - cmin;                           // Texel 1 is selected
    } else {
        cmin = cmax;
        float2 uv1 = float2(uv.x-rr, uv.y);
        cmax += tex2D(texture, uv1);            // Texel 2
        if(cmax >= r) {
            r = r - cmin;                       // Texel 2 is selected
            uv = uv1;
        } else {
            cmin = cmax;
            uv1 = float2(uv.x, uv.y-rr);
            cmax += tex2D(texture, uv1);        // Texel 3
            if(cmax >= r) {
                r = r - cmin;                   // Texel 3 is selected
                uv = uv1;
            } else {                            // Texel 4
                cmin = cmax;
                uv1 = float2(uv.x-rr, uv.y-rr);
                r = r - cmin;                   // Texel 4 is selected
                uv = uv1;
            }
        }
    }
}
```

7.1.2   Update of the Radiance Texture

The points visible from selected shooter $\mathbf{y}$ can be found by placing a hemicube around $\mathbf{y}$, and then using the z-buffer algorithm to identify the visible patches. Since it turns out just at the end, i.e. having processed all patches by the z-buffer algorithm, which points are really visible, the radiance update requires two passes.

The center and the base of the hemicube are set to $\mathbf{y}$ and to the surface at $\mathbf{y}$, respectively. Taking the hemicube center and side faces as the eye and window, respectively, the scene is rendered five times, also computing the z coordinate of the visible points. These values are written into a texture, called the *depth map*.

Another option might be the application of *hemispherical mapping* [26] (Figure 7.4), which would require just a single rendering pass, but it is usually not worth taking because the

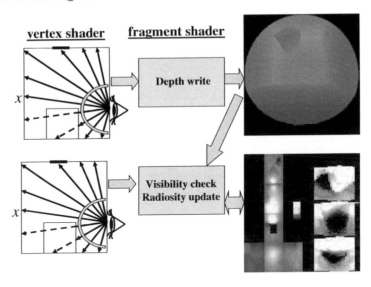

FIGURE 7.4: The two passes of the radiance update. The first pass generates a depth map to identify points visible from the shooter. The second pass transfers radiance to these points.

distortion of hemispherical projection introduces artifacts. The problem is that hemispherical projection is nonlinear, and maps triangles to regions bounded by ellipses. However, the current graphics hardware always assumes that the points leaving the vertex shader are triangle vertices, and generates those pixels that are inside these triangles. This is acceptable only if the triangles are very small, i.e. when the distortion caused by the hemispherical mapping is negligible.

In the second pass we render into the rectangle of the radiance texture. It means that the fragment shader will visit each texel, and updates the stored actual radiance in texture space. The viewport resolution is set to the texture resolution and the render target is the radiance texture. The vertex shader is set to map a point of the unit texture square onto the full viewport. Normalized device space location hPos (which is identical to clipping space location since the fourth homogeneous coordinate is set to 1) is computed from texture coordinates texx of the receiver having position Pos as follows:

```
void RadiosityUpdateVS(
    in  float4 Pos      : POSITION,   // Input receiver position in modeling space
    in  float2 Texx     : TEXCOORD0,  // Receiver patch
    out float4 hPos     : POSITION,   // Clipping space for rendering to texture
    out float4 oTexx    : TEXCOORD0,  // Output receiver patch
    out float3 x        : TEXCOORD1,  // Output receiver position in camera space
    out float3 xnorm    : TEXCOORD2,  // Output normal of the receiver in camera space
```

```
out float4 vch       : TEXCOORD3) // Clipping space for rendering to depth map
{
    oTexx = Texx;        // Texture coordinates identifying the receiver patch
    hPos.x = 2 * Texx.x - 1;    // [0,1]^2 texture space -> [-1,1]^2 clipping space
    hPos.y = 1 - 2 * Texx.y;
    hPos.z = 0;
    hPos.w = 1;
```

The transformation between texture space and normalized device space coordinates is necessary because device space coordinates must be in $[-1, 1]$, while texture coordinates are expected in $[0, 1]$. The origin of the texture space is the upper left corner and the y-axis points downward, so we need to flip the y coordinates.

The vertex shader also transforms input vertex Pos to camera space (x), as well as its normal vector xnorm to compute radiance transfer, determines homogeneous coordinates vch for the location of the point in the depth map.

```
    x     = mul(Pos, WorldView).xyz;      // Receiver in camera space
    xnorm = mul(xnorm, WorldViewIT).xyz;  // Normal of the receiver in camera space
    vch   = mul(Pos, WorldViewProj);      // Receiver in clipping space of the depth map
}
```

The geometric factor depends on the receiver point, thus its accurate evaluation could be implemented by the fragment shader. The fragment shader gets the shooter patch id (Texy) and the probability of having selected this patch (p) as uniform parameters. Recall that this probability is the ratio of the power luminance of the selected texel and the total power luminance of the scene.

```
float4 RadiosityUpdatePS(
    in float4 Texx   : TEXCOORD0, // Receiver patch
    in float3 x      : TEXCOORD1, // Receiver position in camera space
    in float3 xnorm  : TEXCOORD2, // Receiver normal in camera space
    in float4 vch    : TEXCOORD3, // Receiver in clipping space of the depth map
    uniform float2 Texy,          // Shooter patch
    uniform float  p              // Selection probability of the shooter
    ) : COLOR

{
    float3 ytox = normalize(x);           // Direction from y to x
    float   xydist2 = dot(x, x);          // |x - y|^2
    float   cthetax = dot(xnorm, -ytox);  // cos(theta_x)
    if (cthetax < 0) costhetax = 0;
    float3 ynorm = float3(0, 0, 1);
    float   cthetay = ytox.z;             // cos(theta_y)
```

```
if (cthetay < 0) costhetay = 0;
float G = cthetax * cthetay / xydist2;  // Geometry factor
```

Note that we took advantage of the fact that **y** is the eye position of the camera, which is transformed to the origin by the `WorldView` transform, and the normal vector at this point is transformed to the axis *z*.

When a texel of the radiance map is shaded, it is checked whether or not the center of the surface corresponding to this texel is visible from the shooter by comparing the depth values stored in the depth map. The fragment shader code responsible for converting the homogeneous coordinates (`vch`) to the Cartesian coordinates (`vcc`) and computing the visibility indicator is

```
float3 vcc = vch.xyz / vch.w; // Cartesian
vcc.x = (vcc.x + 1) / 2;       // Texture space
vcc.y = (1 - vcc.y) / 2;
float depth = tex2D(depthmap, vcc).r;
float vis = (abs(depth - vcc.z) < EPS);
```

Instead of coding these steps manually, we could use projective depth texturing as well, as discussed in Section 4.1 on *shadow mapping*.

To obtain the radiance transfer from shooter **y** to the processed receiver point **x**, first the radiance of shooter **y** is calculated from its reflected radiance stored in `radmap`, and its emission stored in `emissmap`. Shooter's texture coordinates Texy are passed as uniform parameters:

```
float3 Iy = tex2D(radmap, Texy);     // Reflected radiance of the shooter
float3 Ey = tex2D(emissmap, Texy);   // Emission of the shooter
float3 Ly = Ey + Iy;                 // Radiance of the shooter
```

The new reflected radiance at receiver **x** is obtained from the radiance at **y** multiplying it with visibility indicator `vis` and geometric factor `G` computed before, and dividing by shooter selection probability p passed as a uniform parameter. The emission and the surface area of this texel are read from texture map `emissmap`. Texel luminance (`lum`) at **x** is also computed to allow importance sampling in the subsequent iteration step, and stored in the alpha channel of the reflected radiance.

```
float4 brdfx = tex2D(brdfmap, Texx);  // f_r(x), BRDF
float3 Lx = Ly * G * vis * brdfx / p;  // L(x) radiance estimate
float4 Ex = tex2D(emissmap, Texx);    // E(x) emission
float  Ax = Ex.a;                     // Surface area of the receiver
float3 em  = float3(0.21, 0.39, 0.4); // Weight for luminance computation
float  lum = dot(Ex + Lx, em) * Ax;   // Luminance for importance sampling
return float4(Lx, lum);
}
```

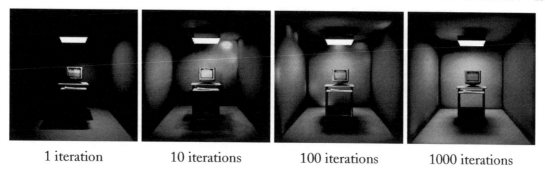

| 1 iteration | 10 iterations | 100 iterations | 1000 iterations |

FIGURE 7.5: Images rendered with stochastic hemicube shooting. All objects are mapped to a single texture map of resolution 128×128, which corresponds to processing $16\,000$ patches.

The implementation has been tested with a room scene of Figure 7.5, and we concluded that a single iteration requires less than 20 ms on an NV6800GT for a few hundred vertices and for 128×128 resolution radiance maps. Using 64×64 resolution radiance maps introduced a minor amount of shadow bleeding, but increased iteration speed by approximately 40%. Since we can expect converged images after 40–80 iterations for normal scenes, this corresponds to 0.5–1 frames per second, without exploiting frame-to-frame coherence. In order to eliminate flickering, we should use the same random number generator in all frames. On the other hand, as in all iterative approaches, frame to frame coherence can be easily exploited. In the case of moving objects, we can take the previous solution as a good guess to start the iteration. This trick not only improves accuracy, but also makes the error of subsequent steps highly correlated, which also helps eliminating flickering.

7.2   PRE-COMPUTED RADIOSITY

In the previous section, we discussed the radiosity algorithm as an on-line, GPU-based method. On the other hand, radiosity is also a popular pre-processing approach, which computes the global illumination solution off-line either on the GPU or on the CPU, and the resulting radiance texture is mapped on the surfaces during on-line rendering. Of course, this is only accurate if the scene geometry, material properties, and light sources are static. To allow both static lighting computed by radiosity and dynamic lights processed on the fly, instead of the reflected radiance, the total irradiance caused by static lights is stored in the generated texture map. This texture map is called the *light map*. Let us revisit the equation of the reflected radiance at a diffuse surface:

$$L^r(\mathbf{x}) = \int_{\Omega'} L^{\text{in}}(\mathbf{x}, \omega') f_r(\mathbf{x}) \cos^+ \theta'_{\mathbf{x}} \, d\omega' = f_r(\mathbf{x}) \int_{\Omega'} L^{\text{in}}(\mathbf{x}, \omega') \cos^+ \theta'_{\mathbf{x}} \, d\omega'$$

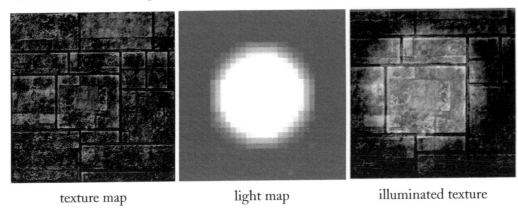

| texture map | light map | illuminated texture |

FIGURE 7.6: Application of a light map.

and let us decompose the incident radiance field L^{in} to static illumination L_s^{in} and to dynamic illumination L_d^{in} usually caused by abstract light sources.

The light map stores the *total irradiance* caused by static light sources either directly or indirectly:

$$I_s(\mathbf{x}) = \int_{\Omega'} L_s^{in}(\mathbf{x}, \omega') \cos^+ \theta_{\mathbf{x}}' \, d\omega'. \qquad (7.4)$$

During the on-line part of rendering the irradiance $I_d(\mathbf{x})$ of dynamic lights is also obtained and added to the light map value. The reflected radiance is then the total irradiance multiplied by the BRDF (Figure 7.6):

$$L^r(\mathbf{x}) = f(\mathbf{x})(I_s(\mathbf{x}) + I_d(\mathbf{x})).$$

Since radiosity simulates light transfer between larger patches and not on the level of surface displacements and normal maps, light maps create a "flat" look. The core of the problem is that in the formula of the total irradiance (equation (7.4)) both the low-frequency incident radiance field, L_s^{in}, and the high-frequency normal vector, $\cos^+ \theta_{\mathbf{x}}'$, are included. The incident radiance changes slowly, thus it can be represented by a lower resolution map. However, in the case of bump, normal, or displacement mapped surfaces, the normal vector changes quickly, which requires a high-resolution texture map. If we take the resolution required by the normal map, then the storage overhead would be high and the radiosity pre-processing would be difficult due to the high number of surface patches. Note that the number of patches is equal to the number of texels in the texture map, and the computational complexity of radiosity algorithms is often quadratic. On the other hand, if we use a lower resolution texture map, then the light map cannot follow the quick changes of the normal vector, resulting in a flat look.

The solution of this problem is the separation of the incident radiance field and the normal vector information. However, unlike the total irradiance, the incident radiance is direction-dependent, thus it cannot be defined by a single value. Let us decompose the directional hemisphere Ω' above the considered surface to finite number of solid angles $\Delta\omega_i'$, $i = 1, \ldots, N$, and we assume that the incident radiance is roughly uniform from directions inside a solid angle. The direction corresponding to the center of solid angle $\Delta\omega_i'$ is denoted by unit direction vector $\mathbf{L}_i$. Using this decomposition, the irradiance caused by the static illumination is:

$$I_s(\mathbf{x}) = \sum_{i=1}^{N} \int_{\Delta\omega_i'} L_s^{\text{in}}(\mathbf{x}, \omega') \cos^+ \theta_{\mathbf{x}}'\, d\omega' \approx \sum_{i=1}^{N} L_s^{\text{in}}(\mathbf{x}, L_i) \Delta\omega_i'(\mathbf{L}_i \cdot \mathbf{N}(\mathbf{x})).$$

We store products $L_s^{\text{in}}(\mathbf{x}, \omega_i')\Delta\omega_i'$ in N light maps. Directions $\mathbf{L}_i$ are fixed for a triangle of a given orientation, so they need not be stored. The computation can be further simplified if dot products $(\mathbf{L}_i \cdot \mathbf{N}(\mathbf{x}))$ are stored in a new "normal map" instead of the original normal map representing normal vector $\mathbf{N}(\mathbf{x})$ [49]. If occlusion factors o_i due to the displaced geometry are also computed in solid angles $\Delta\omega_i'$, then they can be incorporated into the new "normal map", which now stores values $o_i(\mathbf{L}_i \cdot \mathbf{N}(\mathbf{x}))$. Using this, self-shadowing can be cheaply simulated.

Surprisingly, few solid angles can provide appealing results. In [49] the number of solid angles (N) was only three.

7.3   DIFFUSE AND GLOSSY FINAL GATHERING WITH DISTANCE MAPS

A cube map can represent the incoming illumination of a single point in all directions, thus, it is a primary candidate for storing incoming illumination [139]. However, the problem is that a single cube map is an exact representation only for a single point, the *reference point* where the cube map images were taken. To solve this problem, we can store geometric information in cube map texels, for example, the distance between the cube map center and the point visible in a given texel. Such a cube map is called the *distance map* (see also Chapter 6). When the illumination of a point other than the cube map center is computed, this geometric information is used to correct the incoming radiance of the cube map texels.

Let us assume that we use a single cube map that was rendered from reference point $\mathbf{o}$. The goal is to reuse this illumination information for other nearby points as well. To do so, we apply approximations that allow us to factor out those components from the reflected radiance formula (equation (1.7)) which strongly depend on the shaded point $\mathbf{x}$.

In order to estimate the reflected radiance integral, directional domain Ω' is partitioned to solid angles $\Delta\omega_i'$, $i = 1, \ldots, N$, where the radiance is roughly uniform in each domain. After

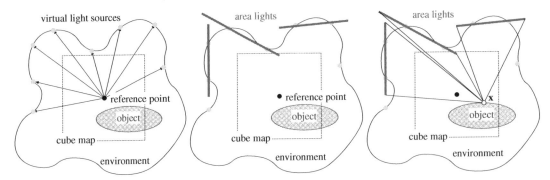

FIGURE 7.7: Diffuse/glossy final gathering. Virtual lights correspond to cube map texels. These point lights are grouped to form large area lights by downsampling the cube map. At shaded point **x**, the illumination of the area lights is computed without visibility tests.

partitioning, the reflected radiance is expressed by the following sum:

$$L^r(\mathbf{x}, \omega) = \sum_{i=1}^{N} \int_{\Delta\omega_i'} L(\mathbf{y}, \omega') f_r(\omega', \mathbf{x}, \omega) \cos^+ \theta_{\mathbf{x}}' \, d\omega'.$$

If $\Delta\omega_i'$ is small, then we can use the following approximation:

$$L^r(\mathbf{x}, \omega) \approx \sum_{i=1}^{N} \tilde{L}^{in}(\Delta y_i) f_r(\omega_i', \mathbf{x}, \omega) \cos^+ \theta_i' \Delta\omega_i', \qquad (7.5)$$

where $\tilde{L}^{in}(\Delta y_i)$ is the *average incoming radiance* from surface Δy_i seen at solid angle $\Delta\omega_i'$. Note that the average incoming radiance is independent of shaded point **x** if the environment is diffuse, and can be supposed to be approximately independent of the shaded point if the environment is moderately glossy. These values can potentially be reused for all shaded points. To exploit this idea, visible surface areas Δy_i need to be identified and their average radiances need to be computed first. These areas are found and the averaging is computed with the help of a cube map placed at reference point **o** in the vicinity of the shaded object. We render the scene from reference point **o** onto the six sides of a cube. In each pixel of these images, the radiance of the visible point and also the distance from the reference point are stored. The pixels of the cube map thus store the radiance and also encode the position of small indirect lights (Figure 7.7).

The small virtual lights are clustered into larger area light sources while averaging their radiance, which corresponds to downsampling the cube map. A pixel of the lower resolution cube map is computed as the average of the included higher resolution pixels. Note that both radiance and distance values are averaged, thus finally we have larger lights having the average radiance of the small lights and placed at their average position. The total area corresponding to a pixel of a lower resolution cube map will be elementary surface Δy_i, and its average radiance is stored in the texel.

The solid angle subtended by a cube map texel of area A can be approximated by the formula of a differential surface of size A:

$$\Delta \omega \approx \frac{A \cos \theta}{d^2},$$ (7.6)

where d is the distance and θ is the angle between the normal vector and the viewing direction. However, this approximation is numerically unstable when distance d gets close to zero. We can obtain a better approximation using the solid angle subtended by a disc. If a disc of area A is perpendicular at its center to the viewing direction, and is at distance d, then it subtends solid angle [46]:

$$\Delta \omega = 2\pi \left(1 - \frac{1}{\sqrt{1 + \frac{A}{d^2 \pi}}} \right).$$ (7.7)

When the surface is not perpendicular, its area should be multiplied with $\cos \theta$.

Supposing that the edge size of the cube map is 2, the area of a texel is $2/M$ where M is the resolution of a single cube map face. Let $\mathbf{L}$ be the vector pointing from the center of the cube map to the texel (Figure 7.8). In this case, the distance is $|\mathbf{L}|$ and $\cos \theta = 1/|\mathbf{L}|$. Thus,

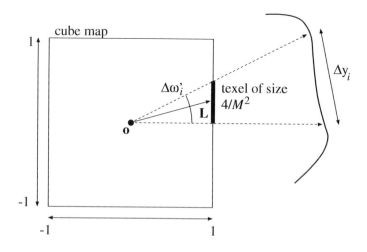

FIGURE 7.8: Solid angle in which a surface is seen through a cube map pixel.

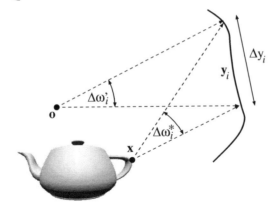

FIGURE 7.9: Notations of the evaluation of subtended solid angles.

the solid angle subtended by a texel is

$$\Delta\omega_i' \approx 2\pi \left(1 - \frac{1}{\sqrt{1 + \frac{4}{M^2 \pi |\mathbf{L}|^3}}} \right).$$

According to equation (7.5), the reflected radiance at the reference point is

$$L^r(\mathbf{o}, \omega) \approx \sum_{i=1}^{N} \tilde{L}^{in}(\Delta y_i) f_r(\omega_i', \mathbf{o}, \omega) \cos^+ \theta_i' \Delta\omega_i'.$$

Let us now consider another point $\mathbf{x}$ close to the reference point $\mathbf{o}$ and evaluate the reflected radiance for point $\mathbf{x}$ while making exactly the same assumption on the surface radiance, i.e. it is constant in area Δy_i:

$$L^r(\mathbf{x}, \omega) \approx \sum_{i=1}^{N} \tilde{L}^{in}(\Delta y_i) f_r(\omega_i^*, \mathbf{x}, \omega) \cos^+ \theta_i^* \Delta\omega_i^*, \qquad (7.8)$$

where $\Delta\omega_i^*$ is the solid angle subtended by Δy_i from $\mathbf{x}$. Unfortunately, the solid angle values can only be obtained directly from the geometry of the cube map if the shaded point is the center of the cube map. In the case of other shaded points, special considerations are needed that are based on the distances from the environment surface.

Solid angle $\Delta\omega_i^*$ is expressed from $\Delta\omega_i'$ using the formula of the solid angle of a disc (equation (7.7)). Assume that the environment surface is not very close compared to the distances of the reference and shaded points, thus the angles between the normal vector at $\mathbf{y}_i$ and illumination directions to $\mathbf{o}$ and $\mathbf{x}$ are similar. In this case, using equation (7.7), we can

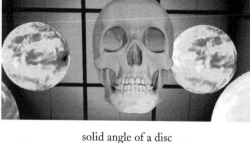

| solid angle of a differential surface | solid angle of a disc |

| visiting all texels | only five texels closest to the reflection direction |

FIGURE 7.10: A diffuse (upper row) and a glossy (lower row) skull rendered with the discussed method. The upper row compares the results of approximations in equations (7.6) and (7.7). The lower row shows the effect of not visiting all texels but only where the BRDF is maximal.

establish the following relationship between $\Delta\omega_i^*$ and $\Delta\omega_i'$:

$$\Delta\omega_i^* \approx 2\pi - \frac{2\pi - \Delta\omega_i'}{\sqrt{(2\pi - \Delta\omega_i')^2 \left(1 - \frac{|\mathbf{o}-\mathbf{y}_i|^2}{|\mathbf{x}-\mathbf{y}_i|^2}\right) + \frac{|\mathbf{o}-\mathbf{y}_i|^2}{|\mathbf{x}-\mathbf{y}_i|^2}}}.$$

Note that the estimation could be made more accurate if the normal vectors were also stored in cube map texels and the cosine angles were evaluated on the fly.

The algorithm first computes an environment cube map from the reference point and stores the radiance and distance values of the points visible in its pixels. We usually generate $6 \times 256 \times 256$ pixel resolution cube maps. Then the cube map is downsampled to have M×M pixel resolution faces (M is 4 or even 2). Texels of the low-resolution cube map represent elementary surfaces Δy_i whose average radiance and distance are stored in the texel. The illumination of these elementary surfaces is reused for an arbitrary point x, as shown by the following HLSL fragment shader program calculating the reflected radiance at this point:

```
float3 RefRadPS (
    float3 N : TEXCOORD0,    // Normal
```

```
        float3 V : TEXCOORD1,      // View direction
        float3 x : TEXCOORD2       // Shader point
        ) : COLOR0
{
        float3 Lr = 0;  // Reflected radiance
        V = normalize(V);
        N = normalize(N);
        for (int X = 0; X < M; X++) // For each texel of a cube map face
            for (int Y = 0; Y < M; Y++) {
                float2 t = float2((X+0.5f)/M, (Y+0.5f)/M);
                float2 l = 2 * t - 1;  // [0,1]->[-1,1]
                Lr += Contrib(x, float3(l.x,l.y, 1), N, V); // Six sides of the cube map
                Lr += Contrib(x, float3(l.x,l.y,-1), N, V);
                Lr += Contrib(x, float3(l.x, 1,l.y), N, V);
                Lr += Contrib(x, float3(l.x,-1,l.y), N, V);
                Lr += Contrib(x, float3( 1,l.x,l.y), N, V);
                Lr += Contrib(x, float3(-1,l.x,l.y), N, V);
        }
        return Lr;
}
```

The Contrib function calculates the contribution of a single texel of the downsampled, low-resolution cube map LREnvMap to the illumination of the shaded point. Arguments x, L, N, and V are the relative position of the shaded point with respect to the reference point, the non-normalized illumination direction pointing to the center of the texel from the reference point, the unit surface normal at the shaded point, and the unit view direction, respectively.

```
float3 Contrib(float3 x, float3 L, float3 N, float3 V) {
    float   l    = length(L);
    float   dw   = 1 / sqrt(1 + 4/(M*M*l*l*l*PI)); // Solid angle of the texel
    float   doy  = texCUBE(LRCubeMap, L).a;
    float   doy2 = doy * doy;
    float3  y    = L / l * doy;
    float   doy_dxy2 = doy2 / dot(y-x, y-x);
    float   dws  = 2*PI - dw / sqrt((dw*dw*(1-doy_dxy2)+doy_dxy2));
    float3 I = normalize(y - x);                  // Illumination direction
    float3 H = normalize(I + V);
    float3 a = kd * max(dot(N,I),0) + ks * pow(max(dot(N,H),0),n);
    float3 Lin  = texCUBE(LRCubeMap, L).rgb;
    return Lin * a * dws;
}
```

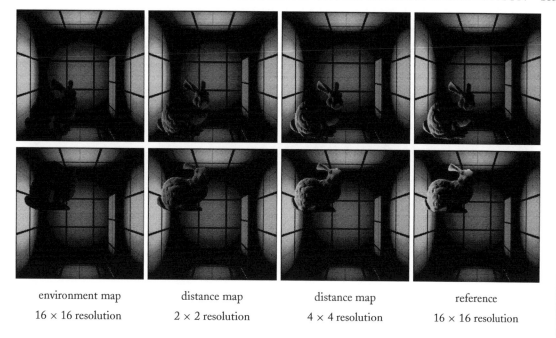

environment map	distance map	distance map	reference
16 × 16 resolution	2 × 2 resolution	4 × 4 resolution	16 × 16 resolution

FIGURE 7.11: Diffuse bunny rendered with the classical environment mapping (left column) and with distance cube maps using different map resolutions.

First, the solid angle subtended by the texel from the reference point is computed and stored in variable dw, then illuminating point y is obtained by looking up the distance value of the cube map. The square distances between the reference point and the illuminating point, and between the shaded point and the illuminating point are put into doy2 and dxy2, respectively. These square distances are used to calculate solid angle dws subtended by the illuminating surface from the shaded point. The Phong–Blinn BRDF is used with diffuse reflectivity kd, glossy reflectivity ks, and shininess n. Illumination direction I and halfway vector H are calculated, and the reflection of the radiance stored in the cube map texel is obtained according to equation (7.8).

In order to demonstrate the results, we took a simple environment of a room containing "fire balls" (Figure 7.10) or a divider face, and put diffuse and glossy objects into this environment. Figure 7.11 shows a diffuse bunny inside the cubic room. The images of the first column are rendered by the traditional environment mapping technique for diffuse materials where a precalculated convolution enables us to determine the irradiance at the reference point with a single lookup. Clearly, these precalculated values cannot deal with the position of the object, thus the bunny looks similar everywhere. The other columns show the results of the distance

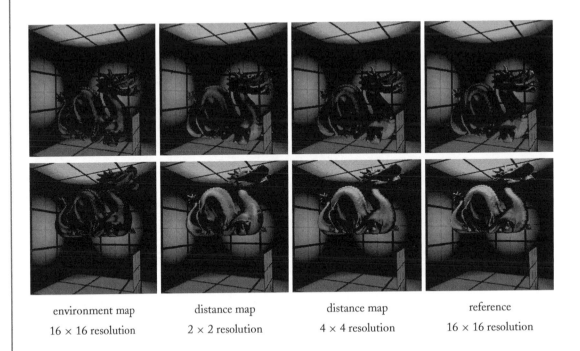

environment map   distance map   distance map   reference

16 × 16 resolution   2 × 2 resolution   4 × 4 resolution   16 × 16 resolution

FIGURE 7.12: Glossy Buddha (the shininess is 5) rendered with the classical environment mapping (left column) and with distance cube maps using different map resolutions.

environment map   distance map   distance map   reference

16 × 16 resolution   2 × 2 resolution   4 × 4 resolution   16 × 16 resolution

FIGURE 7.13: Glossy dragon (the shininess is 5) rendered with the classical environment mapping (left column) and with distance cube maps using different map resolutions.

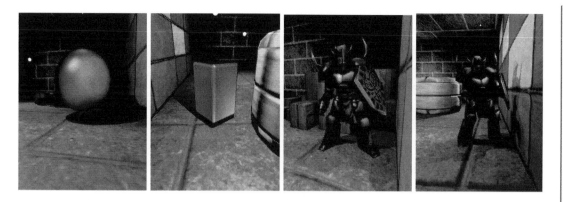

FIGURE 7.14: Glossy objects and a knight rendered with the distance map-based final gathering algorithms.

map-based method using different resolution cube maps. Note that even with extremely low resolution (2×2) we get images similar to the large-resolution reference.

Figures 7.12 and 7.13, respectively, show a glossy Buddha and a dragon inside a room. The first column presents the traditional environment mapping technique while the other three columns present the results of the localized algorithm. Similarly to the diffuse case, even cube map resolution of 2×2 produced pleasing results. Images of games using these techniques are shown in Figures 7.14 and 7.15.

FIGURE 7.15: Scientist with indirect illumination obtained by the distance map-based method (left) and by the classic local illumination method (right) for comparison.

CHAPTER 8

Pre-computation Aided Global Illumination

Global illumination algorithms should integrate the contribution of all light paths connecting the eye and the light sources via scattering points. If the scene is static, that is, its geometry and material properties do not change in time, then these light paths remain the same except for the first light and the last viewing rays, which might be modified due to moving lights and camera. Pre-computation aided approaches pre-compute the effects of the static parts of the light paths, and evaluate just the light and viewing rays run-time.

Let us consider a light beam of radiance $L^{in}(\mathbf{y}, \omega_i)$ approaching *entry point* $\mathbf{y}$ on the object surface from direction ω_i (Figure 8.1). Note that this light beam may be occluded from point $\mathbf{y}$, prohibiting the birth of the path at $\mathbf{y}$, and making the contribution zero. If the light beam really reaches $\mathbf{y}$, then it may be reflected, refracted even several times, and finally it may arrive at the eye from *exit point* $\mathbf{x}$ with viewing direction ω_o. Entry and exit points can be connected by infinitely many possible paths, from which a finite number can be generated by the Monte Carlo simulation, for example, by random walks or iterative techniques [134, 35].

The total contribution of the set of the paths arriving at $\mathbf{y}$ from direction ω_i to the visible radiance at $\mathbf{x}$ and direction ω_o is

$$L^{in}(\mathbf{y}, \omega_i)T(\mathbf{y}, \omega_i \to \mathbf{x}, \omega_o),$$

where $T(\mathbf{y}, \omega_i \to \mathbf{x}, \omega_o)$ is the *transfer function*. The total visible radiance $L^{out}(\mathbf{x}, \omega_o)$ is the sum of the contributions of all paths entering the scene at all possible entry points and from all possible directions, thus it can be expressed by the following double integral:

$$L^{out}(\mathbf{x}, \omega_o) = \int_S \int_\Omega L^{in}(\mathbf{y}, \omega_i)T(\mathbf{y}, \omega_i \to \mathbf{x}, \omega_o)\, d\omega_i dy \qquad (8.1)$$

where S is the set of surface points and Ω is the set of directions. We can conclude that in the general case the output radiance is obtained as a double integral of the product of a four-variate

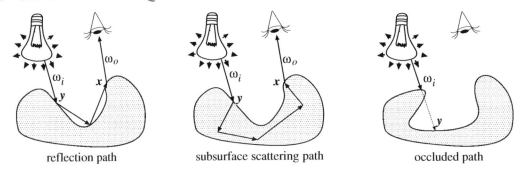

reflection path   subsurface scattering path   occluded path

FIGURE 8.1: Light paths sharing the same light and viewing rays.

transfer function and a two-variate incoming radiance function. This is rather complicated, but fortunately there are a couple of important simpler cases:

1. *Diffuse surfaces* have direction independent outgoing radiance, thus both exit radiance L^{out} and transfer function T become independent of ω_o:

$$L^{\text{out}}(\mathbf{x}) = \int\limits_{S} \int\limits_{\Omega} L^{\text{in}}(\mathbf{y}, \omega_i) T(\mathbf{y}, \omega_i \to \mathbf{x})\, d\omega_i dy.$$

The number of variables is reduced by one in the transfer function.

2. In the case of *directional lights* and *environment map* lighting (also called *image-based lighting* or *sky illumination*), the incoming radiance L^{in} gets independent of entry point $\mathbf{y}$, thus we can write

$$L^{\text{out}}(\mathbf{x}, \omega_o) = \int\limits_{S} \int\limits_{\Omega} L^{\text{in}}(\omega_i) T(\mathbf{y}, \omega_i \to \mathbf{x}, \omega_o)\, d\omega_i dy$$

$$= \int\limits_{\Omega} L^{\text{in}}(\omega_i) T^{\text{env}}(\omega_i \to \mathbf{x}, \omega_o)\, d\omega_i$$

where

$$T^{\text{env}}(\omega_i \to \mathbf{x}, \omega_o) = \int\limits_{S} T(\mathbf{y}, \omega_i \to \mathbf{x}, \omega_o)\, dy$$

is the total transfer from direction ω_i to exit point $\mathbf{x}$. Again, we can observe the reduction of the number of independent variables by one in the transfer function.

3. Diffuse surfaces illuminated by directional lights or environment maps have an even simpler formula where the number of variables of the transfer function is

reduced to two:

$$L^{\text{out}}(\mathbf{x}) = \int_\Omega L^{\text{in}}(\omega_i) T^{\text{env}}(\omega_i \to \mathbf{x}) \, d\omega_i.$$

The transfer function depends just on the geometry and the material properties of objects, and is independent of the actual lighting, thus it can be pre-computed for certain entry and exit points.

The problem is then how we can determine and represent a function defined for infinitely many points and directions, using finite amount of data. There are two straightforward solutions, *sampling* and the *finite-element method*.

8.1   SAMPLING

Sampling does not aim at representing the function everywhere in its continuous domain, but only at finite number of sample points.

For example, we usually do not need the output radiance everywhere only at sample points $\mathbf{x}_1, \ldots, \mathbf{x}_K$. Such sample points can be the vertices of a highly tessellated mesh (*per–vertex approach*), or the points corresponding to the texel centers (*per–pixel approach*). It might happen that non-sample points may also turn out to be visible, and their radiance is needed. In such cases, *interpolation* can be applied taking the neighboring sample points. Linear interpolation is directly supported by the graphics hardware. When sample points are the vertices, Gouraud shading can be applied, or when the points correspond to the texel centers, bi-linear texture filtering executes the required interpolation.

8.2   FINITE-ELEMENT METHOD

According to the concept of the *finite-element method* (see Section 1.6), the transfer function is approximated by a finite function series form:

$$T(\mathbf{y}, \omega_i \to \mathbf{x}, \omega_o) \approx \sum_{n=1}^{N} \sum_{m=1}^{M} \sum_{k=1}^{K} \sum_{l=1}^{L} s_n(\mathbf{y}) d_m(\omega_i) S_k(\mathbf{x}) D_l(\omega_o) T_{nmkl},$$

where $s_n(\mathbf{x})$ and $S_k(\mathbf{x})$ are pre-defined spatial *basis functions*, and $d_m(\omega_i)$ and $D_l(\omega_i)$ are pre-defined directional basis functions. Since the basis functions are known, the transfer function at any $\mathbf{y}, \omega_i, \mathbf{x}, \omega_o$ is determined by coefficients T_{nmkl}. In the general case, we have $N \times M \times K \times L$ number of such coefficients.

Having selected the basis functions, the coefficients are obtained evaluating the following integrals, using usually the Monte Carlo quadrature:

$$T_{nmkl} = \int_S \int_\Omega \int_S \int_\Omega T(\mathbf{y}, \omega_i \to \mathbf{x}, \omega_o) \tilde{s}_n(\mathbf{y}) \tilde{d}_m(\omega_i) \tilde{S}_k(\mathbf{x}) \tilde{D}_l(\omega_o)\, d\omega_i dy d\omega_o dx,$$

where $\tilde{s}_n(\mathbf{y}), \tilde{d}_m(\omega_i), \tilde{S}_k(\mathbf{x}), \tilde{D}_l(\omega_o)$ are *adjoints* of basis functions $s_n(\mathbf{y}), d_m(\omega_i), S_k(\mathbf{x}), D_l(\omega_o)$, respectively. Basis functions $\tilde{s}_1, \ldots, \tilde{s}_N$ are said to be adjoints of $s_1, \ldots, s_N$ if the following condition holds:

$$\int_S \tilde{s}_i(\mathbf{y}) s_j(\mathbf{y})\, d y = \delta_{ij},$$

where $\delta_{ij} = 0$ if $i \neq j$ (*orthogonality*) and $\delta_{ii} = 1$ (*normalization*). A similar definition holds for directional basis functions as well.

Piecewise constant basis functions are orthogonal to each other. To guarantee that the normalization constraint is also met, that is $\int_S \tilde{s}_i(\mathbf{y}) s_i(\mathbf{y})\, d y = 1$, the adjoint constant basis functions are equal to the reciprocal of the size of their respective domains. Real spherical harmonics and Haar wavelets are also self-adjoint.

Substituting the finite-element approximation into the exit radiance, we can write

$$L^{out}(\mathbf{x}, \omega_o) = \sum_{k=1}^{K} \sum_{l=1}^{L} \sum_{n=1}^{N} \sum_{m=1}^{M} S_k(\mathbf{x}) D_l(\omega_o) T_{nmkl} \int_S \int_\Omega L^{in}(\mathbf{y}, \omega_i) s_n(\mathbf{y}) d_m(\omega_i)\, d\omega_i dy.$$

(8.2)

Note that after pre-computing coefficients T_{nmkl} just a low-dimensional integral needs to be evaluated. This computation can further be speeded up if illumination function $L^{in}(\mathbf{y}, \omega_i)$ is also approximated by a finite-element form

$$L^{in}(\mathbf{y}, \omega_i) \approx \sum_{n'=1}^{N} \sum_{m'=1}^{M} \tilde{s}_{n'}(\mathbf{y}) \tilde{d}_{m'}(\omega_i) L_{n'm'}$$

where the basis functions $\tilde{s}_{n'}$ and $\tilde{d}_{m'}$ are the *adjoints* of s_n and d_m, respectively.

Substituting the finite-element approximation of the incoming radiance into the integral of equation (8.2), we can write

$$\int_S \int_\Omega L^{in}(\mathbf{y},\omega_i)s_n(\mathbf{y})d_m(\omega_i)\,d\omega_i dy \approx$$

$$\int_S \int_\Omega \sum_{n'=1}^N \sum_{m'=1}^M \tilde{s}_{n'}(\mathbf{y})\tilde{d}_{m'}(\omega_i)s_n(\mathbf{y})d_m(\omega_i)L_{n'm'}\,d\omega_i dy = L_{nm}$$

since only those terms are nonzero where $n' = n$ and $m' = m$. Thus, the exit radiance is

$$L^{out}(\mathbf{x},\omega_o) = \sum_{k=1}^K \sum_{l=1}^L \sum_{n=1}^N \sum_{m=1}^M S_k(\mathbf{x})D_l(\omega_o)T_{nmkl}L_{nm}. \qquad (8.3)$$

Different pre-computation aided real-time global illumination algorithms can be classified according to where they use sampling or finite-element methods, and to the considered special cases. For example, all methods published so far work with sample points and apply linear interpolation to handle the positional variation of exit point $\mathbf{x}$. On the other hand, the incoming direction is attacked both by different types of finite-element basis functions and by sampling.

8.2.1   Compression of Transfer Coefficients

The fundamental problem of pre-computation aided global illumination algorithms is that they require considerable memory space to store the transfer function coefficients. To cope with the memory requirements, data compression methods should be applied. A popular approach is the *principal component analysis (PCA)* [127]. Values T_{nmkl} can be imagined as K number of points $\mathbf{T}^1 = [T_{nm1l}], \ldots, \mathbf{T}^K = [T_{nmKl}]$ in an $N \times M \times L$-dimensional space. To compress this data, we find a subspace (a "hyperplane") in the high-dimensional space and project points $\mathbf{T}^i$ onto this subspace. Since the subspace has lower dimensionality, the projected points can be expressed by fewer coordinates, which results in data compression.

Let us denote the origin of the subspace by $\mathbf{M}$ and the unit length, orthogonal basis vectors of the low-dimensional subspace by $\mathbf{B}_1, \ldots, \mathbf{B}_D$. Projecting into this subspace means the following approximation:

$$\mathbf{T}^i \approx \tilde{\mathbf{T}}^i = \mathbf{M} + w_1^i \mathbf{B}_1 + \cdots + w_D^i \mathbf{B}_D,$$

where $w_1^i, w_2^i, \ldots, w_D^i$ are the coordinates of the projected point in the subspace coordinate system. Orthogonal projection results in the following approximating coordinates:

$$w_j^i = (\mathbf{T}^i - \mathbf{M}) \cdot \mathbf{B}_j.$$

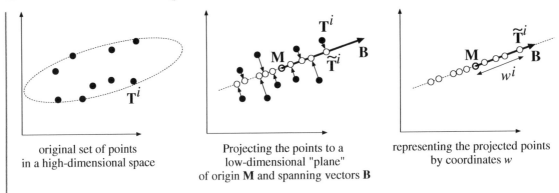

| original set of points in a high-dimensional space | Projecting the points to a low-dimensional "plane" of origin **M** and spanning vectors **B** | representing the projected points by coordinates w |

FIGURE 8.2: The basic idea of *principal component analysis*. Points in a high-dimensional (two-dimensional in the figure) space are projected onto a lower, D-dimensional subspace ($D = 1$ in the figure, thus the subspace is a line), and are given by coordinates in this low-dimensional subspace. To define these coordinates, we need a new origin **M** and basis vectors $\mathbf{B}_1, \ldots, \mathbf{B}_D$ in the lower-dimensional subspace. The origin can be the mean of original sample points. In the example of the figure there is only one basis vector, which is the direction vector of the line.

Of course, origin **M** and basis vectors $\mathbf{B}_1, \ldots, \mathbf{B}_D$ must be selected to minimize the total approximation error, i.e. the sum of the square distances between the original and projected points. As can be shown, the error is minimal if the origin is the mean of the original data points

$$\mathbf{M} = \frac{1}{K} \sum_{k=1}^{K} \mathbf{T}^k,$$

and the basis vectors are the eigenvectors corresponding to the largest D eigenvalues of *covariance matrix*

$$\sum_{k=1}^{K} (\mathbf{T}^k - \mathbf{M})^T \cdot (\mathbf{T}^k - \mathbf{M}),$$

where superscript T denotes transpose operation (a row vector is turned to be a column vector). Intuitively, this error minimization process corresponds to finding a minimal ellipsoid that encloses the original points, and obtaining the center of this ellipsoid as **M** and the longest axes of the ellipsoid to define the required hyperplane. Indeed, if the remaining axes of the ellipsoid are small, then the ellipsoid is flat and can be well approximated by the plane of other axes.

8.3   PRE-COMPUTED RADIANCE TRANSFER

The classical *pre-computed radiance transfer* (*PRT*) [128, 50] assumes that the illumination comes from directional lights or from environment lighting and the surfaces are diffuse, reducing the variables of the transfer function to exit point $\mathbf{x}$ and incoming direction ω_i. The exit radiance is obtained just at sampling points $\mathbf{x}_k$. Sampling points can be either the vertices of the triangle mesh (*per–vertex PRT*) or those points that correspond to the texel centers of a texture atlas (*per-pixel PRT*). The remaining directional dependence on the incoming illumination is represented by directional basis functions $d_1(\omega), \ldots, d_M(\omega)$:

$$T^{\mathrm{env}}(\omega_i \to \mathbf{x}_k) \approx \sum_{m=1}^{M} T_m d_m(\omega_i).$$

The environment illumination, including directional lights, on the other hand, are expressed by the adjoint basis functions:

$$L^{\mathrm{in}}(\omega_i) \approx \sum_{m'=1}^{M} L_{m'} \tilde{d}_{m'}(\omega_i).$$

In this special case, the output radiance at the sample point $\mathbf{x}_k$ is

$$L^{\mathrm{out}}(\mathbf{x}_k) = \int_{\Omega} L^{\mathrm{in}}(\omega_i) T^{\mathrm{env}}(\omega_i \to \mathbf{x}_k)\, d\omega_i \approx$$

$$\sum_{m=1}^{M} T_m \sum_{m'=1}^{M} L_{m'} \int_{\Omega} \tilde{d}_{m'}(\omega_i) d_m(\omega_i)\, d\omega_i = \sum_{m=1}^{M} T_m L_m. \tag{8.4}$$

If $\mathbf{T} = [T_1, \ldots, T_M]$ and $\mathbf{L} = [L_1, \ldots, L_M]$ are interpreted as M-dimensional vectors, the result becomes an M-dimensional dot product. This is why this approach is often referred to as the *dot product illumination*. Note that this formula is able to cope with translucent materials (subsurface scattering) [87] as well as opaque objects.

   If the transfer function samples are compressed, the reflected radiance can be obtained as follows:

$$L^{\mathrm{out}}(\mathbf{x}_i) \approx \mathbf{L} \cdot \mathbf{T}(\mathbf{x}_i) \approx \mathbf{L} \cdot \left(\mathbf{M} + w_1^i \mathbf{B}_1 + \cdots + w_n^i \mathbf{B}_D \right) = \mathbf{L} \cdot \mathbf{M} + \sum_{d=1}^{D} w_d^i (\mathbf{L} \cdot \mathbf{B}_d).$$

   If there are many original points, we usually cannot expect them to be close to a hyperplane, thus this approach may have a large error. The error, however, can be significantly decreased if we do not intend to find a subspace for all points at once, but cluster points first in a way that points of a cluster are roughly in a hyperplane, then carry out PCA separately for each cluster. This process is called *Clustered Principal Component Analysis (CPCA)*. We can group those points that are close and have similar normal vectors, or apply the *K-means algorithm*[1][73].

Let us denote the means and the basis vectors of cluster c by $\mathbf{M}_c$, and $\mathbf{B}_1^c, \ldots, \mathbf{B}_D^c$, respectively. Each sample point $\mathbf{x}_i$ belongs to exactly one cluster. If point $\mathbf{x}_i$ belongs to cluster c, then its reflected radiance is

$$L^{\text{out}}(\mathbf{x}_i) \approx \mathbf{L} \cdot \mathbf{T}(\mathbf{x}_i) \approx \mathbf{L} \cdot \mathbf{M}^c + \sum_{d=1}^{D} w_d^i (\mathbf{L} \cdot \mathbf{B}_d^c). \qquad (8.5)$$

Implementing the clustering algorithm, then computing the means and the eigenvectors of the clusters, and finally projecting the points of the clusters to subspaces are non-trivial tasks. Fortunately, functions exist in Direct3D 9 that can do these jobs.

When Sloan introduced PRT, he proposed the application of *real spherical harmonics* basis functions in the directional domain [128]. Real spherical harmonics extend the *associated Legendre polynomials* to the directional sphere. So first we should get acquainted with these polynomials. Associated Legendre polynomials form a basis function system in interval $[-1, 1]$ (associated here means that we need scalar valued functions rather than the complex Legendre polynomials). A particular associated Legendre polynomial $P_l^m(x)$ is selected by two parameters l, m, where $l = 0, 1, \ldots$, and $m = 0, 1, \ldots, l$. Polynomials having the same l parameter are said to form a *band*. The sequence of associated Legendre polynomials can be imagined as a triangular grid, where each row corresponds to a band:

$$P_0^0(x)$$
$$P_1^0(x), \ P_1^1(x)$$
$$P_2^0(x), \ P_2^1(x), \ P_2^2(x)$$
$$\cdots$$

This means that we have $n(n+1)/2$ basis functions in an n-band system.

Associated Legendre polynomials can be defined by recurrence relations (similarly to the Cox-deBoor formula of NURBS basis functions). We need three rules. The first rule increases the band:

$$P_l^m(x) = x \frac{2l-1}{l-m} P_{l-1}^m(x) - \frac{l+m-1}{l-m} P_{l-2}^m(x), \quad \text{if } l > m.$$

---

[1]The *K-means* algorithm starts the clustering by randomly classifying points in K initial clusters. Then the mean of each cluster is computed and the distances of each point and each cluster center are found. If a point is closer to the mean of another cluster than to the mean of its own cluster, then the point is moved to the other cluster. Then cluster means are re-computed, and these steps are repeated iteratively.

To start the recurrence, we can use the second rule:

$$P_m^m(x) = (-1)^m (2l - 1)!! (1 - x^2)^{m/2},$$

where $(2l - 1)!!$ is the product of all odd integers less than or equal to $2l - 1$ (*double factorial*). The last rule also allows to lift a term to a higher band:

$$P_{m+1}^m(x) = x(2m + 1) P_m^m(x).$$

The C function computing the value of associated Legendre polynomial $P_l^m(x)$ is [50]:

```
double P(int l, int m, double x) {
    double pmm = 1;
    if (m > 0) {
        double somx2 = sqrt((1-x) * (1+x));
        double fact = 1;
        for(int i=1; i<=m; i++) {
            pmm *= (-fact) * somx2;
            fact += 2;
        }
    }
    if (l == m) return pmm;
    double pmmp1 = x * (2*m+1) * pmm;
    if (l == m+1) return pmmp1;
    double pll = 0;
    for(int ll=m+2; ll<=l; ll++) {
        pll = ( (2*ll-1)*x*pmmp1-(ll+m-1)*pmm ) / (ll-m);
        pmm = pmmp1;
        pmmp1 = pll;
    }
    return pll;
}
```

Now let us extend the associated Legendre polynomials defined in $[-1, 1]$ to the directional domain. The result of this process is the collection of *real spherical harmonics*. Real spherical harmonics also have two parameters l, m, but now m can take values from $-l$ to l. When m is negative, we use the Legendre polynomial defined by $|m|$. The definition of real spherical basis (*SH*) functions is

$$
\begin{aligned}
\text{if } m > 0 \quad & Y_l^m(\theta, \phi) = \sqrt{2} K_l^m P_l^m(\cos\theta) \cos(m\phi), \\
\text{if } m < 0 \quad & Y_l^m(\theta, \phi) = \sqrt{2} K_l^m P_l^{-m}(\cos\theta) \sin(-m\phi), \\
\text{if } m = 0 \quad & Y_l^m(\theta, \phi) = K_l^0 P_l^0(\cos\theta),
\end{aligned}
$$

where

$$K_l^m = \sqrt{\frac{2l+1}{4\pi}\frac{(k-|m|)!}{(l+|m|)!}}$$

is a normalization constant. The code for evaluating an SH function is [50]:

```
double K(int l, int m) { // normalization constant for SH function
    return sqrt( ((2*l+1)*factorial(l-m)) / (4*M_PI*factorial(l+m)) );
}


double SH(int l, int m, double theta, double phi) { // SH basis function
    // l is the band in the range of [0..n], m in the range [-l..l]
    // theta in the range [0..Pi], phi in the range [0..2*Pi]
    if (m == 0)     return K(l,0) * P(l,m,cos(theta));
    else if (m > 0) return sqrt(2.0) * K(l, m) * cos(m*phi) * P(l,m,cos(theta));
    else            return sqrt(2.0) * K(l,-m) * sin(-m*phi) * P(l,-m,cos(theta));
}
```

The application of these spherical basis functions is quite straightforward. If we need N basis functions, we find number of bands n so that $n(n+1)/2 \approx N$, and SH functions for all bands $l = 0 \ldots n$. In practice n is quite small, thus instead of using the general formulae and code, it is worth precomputing these basis functions in algebraic form by hand or with the help of a mathematical software, e.g. Maple. The following program obtains the first 15 basis function values and puts them into array ylm[i] using indexing $i = l(l+1) + m$:

```
void SHEval(float theta, float phi) {
    // Convert spherical angles to Cartesian coordinates
    float x = cos(phi) * sin(theta), y = sin(phi) * sin(theta), z = cos(theta);
    float x2 = x * x, y2 = y * y, z2 = z * z;

    ylm[ 0] = 0.282095;                    // l= 0, m= 0
    ylm[ 1] = 0.488603 * y;                // l= 1, m= -1
    ylm[ 2] = 0.488603 * z;                // l= 1, m= 0
    ylm[ 3] = 0.488603 * x;                // l= 1, m= 1
    ylm[ 4] = 1.092548 * x * y;            // l= 2, m= -2
    ylm[ 5] = 1.092548 * y * z;            // l= 2, m= -1
    ylm[ 6] = 0.315392 * (3 * z2 - 1);     // l= 2, m= 0
    ylm[ 7] = 1.092548 * x * z;            // l= 2, m= 1
    ylm[ 8] = 0.546274 * (x2 - y2);        // l= 2, m= 2
    ylm[ 9] = 0.590044 * y * (3 * x2 - y2);  // l= 3, m= -3
    ylm[10] = 1.445306 * 2 * x * y * z;    // l= 3, m= -2
```

```
ylm[11] = 0.457046 * y * (5 * z2 - 1);    // l= 3, m= -1
ylm[12] = 0.373176 * z * (5 * z2 -3);     // l= 3, m= 0
ylm[13] = 0.457046 * x * (5 * z2 -1);     // l= 3, m= 1
ylm[14] = 1.445306 * z * (x2 - y2);       // l= 3, m= 2
ylm[15] = 0.590044 * x * (x2 - 3 * y2);   // l= 3, m= 3
}
```

SH basis functions are orthogonal, i.e. the adjoint basis functions are the same spherical harmonics. A unique property of SH functions is that they are rotationally invariant. This means that when the object (or the environment map) is rotated, then the process needs not be started from scratch [50].

Since the introduction of PRT it has been extended in various directions.

All-frequency PRT. While short spherical harmonics series are good to approximate smooth functions, they fail to accurately represent sharp changes. To solve this problem, Ng [95] replaced spherical harmonics by *Haar-wavelets* (Figure 1.12) to allow high-frequency environment maps. However, working with wavelets we lose the rotational invariance of spherical harmonics [154].

Non-diffuse PRT. If non-diffuse scenes are also considered under environment map illumination, the output radiance formula gets a little more complex:

$$L^{\text{out}}(\mathbf{x}_k, \omega_o) = \sum_{l=1}^{L} \sum_{m=1}^{M} D_l(\omega_o) T_{ml} L_m, \tag{8.6}$$

which means that the radiance vector is multiplied by a matrix [76, 86].

Localized PRT. Pre-computed radiance transfer assumed an infinitely far illumination source (directional or environment map illumination), which eliminated the dependence on entry point **y**. To cope with light sources at finite distance, the dependence of entry point **y** should also be handled. Such approaches are called *local methods*. In [2], a first-order Taylor approximation was used to consider mid-range illumination. Kristensen et al. [83] discretized the incident lighting into a set of localized (i.e. point) lights. Having a point light, we should integrate only in the domain of incoming directions since for a point light, the direction unambiguously determines entry point **y**.

8.3.1   Direct3D Support of the Diffuse Pre-computed Radiance Transfer

PRT has become an integral part of Direct3D 9, which implements both the per-vertex and per-pixel approaches. The process of the application of the built-in functions during

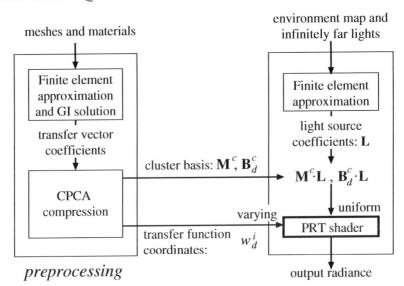

FIGURE 8.3: The simplified block diagram of Direct3D PRT. The transfer function compression results in cluster centers $\mathbf{M}^c$ and cluster basis vector $\mathbf{B}^c_d$, while lighting processing results in vector $\mathbf{L}$. In order to evaluate equation (8.5), the CPU computes scalar products $\mathbf{L} \cdot \mathbf{M}^c$ and $\mathbf{L} \cdot \mathbf{B}^c_d$ and passes them to the shader as uniform parameters. The varying parameters of the mesh vertices are transfer vector coordinates w^d_i.

pre-processing is

1. Preprocessing of the transfer functions, when we have to calculate the transfer function values $\mathbf{T}^i$ at each vertex of the highly tessellated mesh.

2. Compression of the transfer function values using CPCA.

3. Preprocessing of the environment lighting resulting in $\mathbf{L} = [L_0, \ldots, L_N]$.

We have to calculate outgoing radiance as the sum of the dot products of the transfer functions and environment lighting as given by equation (8.5). In this equation, dot products $\mathbf{L} \cdot \mathbf{M}^c$ and $\mathbf{L} \cdot \mathbf{B}^c_d$ are independent of the vertices, thus they can be computed on the CPU and passed to the GPU as a uniform parameter array. These dot products are obtained for each cluster $c = 1, \ldots, C$, dimension $d = 1, \ldots, D$, and for each channel of red, green, and blue, thus the array contains $3C(D+1)$ floats. Since GPUs work with `float4` data elements these constants can also be stored in $C(1 + 3(D+1)/4)$ number of `float4` registers (`Dots` in the implementation). The array offset of the data belonging to cluster k is also passed as an input parameter.

The GPU is responsible for multiplying with weights w_i^d, i.e. the per-vertex data, and adding the results together. The weight and the cluster ID of a sample point are passed as texture coordinates. In the following function the number of clusters C is denoted by NCLUSTERS and the number of dimensions $D + 1$ by NPCA:

```
float4 Dots[NCLUSTERS*(1+3*(NPCA/4))]; // Dot products computed on the CPU

float4 GetPRTDiffuse( int k, float4 w[NPCA/4] ) {
    float4 col = dots[k]; // (M[k] dot L)
    float4 R = float4(0,0,0,0), G = float4(0,0,0,0), B = float4(0,0,0,0);
    // Add (B[k][j] dot L) for each j = 1.. NPCA
    // Since 4 values are added at a time, the loop is iterated until j < NPCA/4
    for (int j=0; j < (NPCA/4); j++) {
        R += w[j] * dots[k+1+(NUM_PCA/4)*0+j];
        G += w[j] * dots[k+1+(NUM_PCA/4)*1+j];
        B += w[j] * dots[k+1+(NUM_PCA/4)*2+j];
    }
    // Sum the elements of 4D vectors
    col.r += dot(R, 1); col.g += dot(G, 1); col.b += dot(B, 1);
    return col;
}
```

A possible per-vertex implementation calls this function from the vertex shader to obtain the vertex color. The fragment shader takes the interpolated color and assigns it to the fragment.

```
void PRTDiffuseVS(
    in  float4 Pos           : POSITION,
    in  int    k             : BLENDWEIGHT, // cluster
    in  float4 w[NUM_PCA/4]  : BLENDWEIGHT1 // weights
    out float4 hPos          : POSITION;
    out float4 Color         : COLOR0;
) {
    hPos = mul(Pos, WorldViewProj);
    Color = GetPRTDiffuse( k, w );
}

float4 StandardPS( in Color : COLOR0 ) : COLOR0 { return Color; }
```

Figures 8.4 and 8.5 show diffuse scenes rendered with PRT. The left of Figure 8.4 has been computed just with single bounces, thus this image is the local illumination solution of image-based lighting. The right of Figure 8.4, on the other hand, was computed simulating

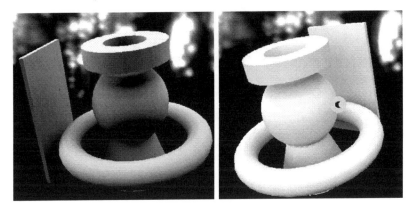

FIGURE 8.4: A diffuse object rendered with PRT. In the left image only single light bounces were approximated with 1024 rays, while the right image contains light paths of maximum length 6. The pre-processing times were 12 s and 173 s, respectively. The resolution of the environment map is 256 × 256. The order of SH approximation is 6.

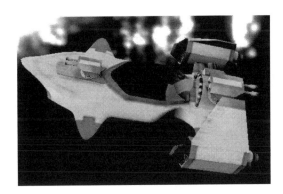

FIGURE 8.5: A diffuse object rendered with PRT. Only single light bounces were approximated with 1024 rays. The resolution of the environment map is 256 × 256. The order of SH approximation is 6. Pre-processing time was 723 s.

six bounces of the light. After preprocessing the camera can be animated at 466 FPS in both cases.

8.4   LIGHT PATH MAP

Pre-computed radiance transfer approaches use finite-element representation for all variables except for exit point **x**, which is handled by sampling. However, we could handle other variables

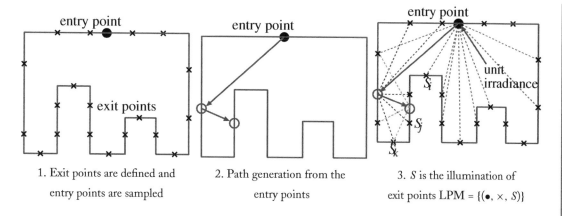

1. Exit points are defined and entry points are sampled

2. Path generation from the entry points

3. S is the illumination of exit points LPM = {(•, ×, S)}

FIGURE 8.6: Overview of the preprocessing phase of the light path map method. Entry points are depicted by •, and exit points by ×. The LPM is a collection of (entry point •, exit point ×, illumination S_k) triplets, called items.

with sampling as well, and the collection of sampled data could be stored in *maps*. If we use sampling, the transfer coefficients are computed for sampled entry–exit point pairs, and stored in textures. The discussed method [132] consists of a preprocessing step and a fast rendering step.

The preprocessing step determines the indirect illumination capabilities of the static scene. This information is computed for finite number of *exit points* on the surface, and we use interpolation for other points. Exit points are depicted by symbol × in Figure 8.6. Exit points can be defined as points corresponding to the texel centers of the texture atlas of the surface.

The first step of preprocessing is the generation of certain number of *entry points* on the surface. These entry points will be samples of first hits of the light emitted by moving light sources. During preprocessing we usually have no specific information about the position and the intensity of the animated light sources, thus entry points are sampled from a uniform distribution, and unit irradiance is assumed at these sample points. Entry points are depicted by symbol • in Figure 8.6. Entry points are used as the start of a given number of random light paths.

The visited points of the generated paths are considered as *virtual light sources* and are connected to all those exit points that are visible from them. In this way, we obtain a lot of paths originating at an entry point and arriving at one of the exit points.

The contribution of a path divided by the probability of the path generation is a Monte Carlo estimate of the indirect illumination. The average of the Monte Carlo estimates of paths associated with the same entry point and exit point pair is stored. We call this data structure the *light path map*, or *LPM* for short. Thus, an LPM contains *items* corresponding to groups

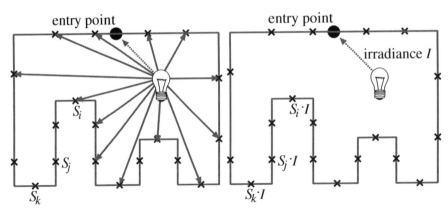

1. Direct illumination + entry point visibility   2. Weighting irradiance I with items S

FIGURE 8.7: Overview of the rendering phase of the light path map method. The irradiance of the entry points are computed, from which the radiance of the exit points is obtained by weighting according to the LPM.

of paths sharing the same entry and exit points. Items that belong to the same entry point constitute an LPM *pane*.

During real-time rendering, LPM is taken advantage to speed up the global illumination calculation. The lights and the camera are placed in the virtual world (Figure 8.7). The direct illumination effects are computed by standard techniques, which usually include some shadow algorithm to identify those points that are visible from the light source. LPM can be used to add the indirect illumination. This step requires visibility information between the light sources and entry points, but this is already available in the shadow maps obtained during direct illumination computation [29].

An LPM pane stores the indirect illumination computed for the case when the respective entry point has unit irradiance. During the rendering phase, however, we have to adapt to a different lighting environment. The LPM panes associated with entry points should be weighted in order to make them reflect the actual lighting situation. Doing this for every entry point and adding up the results, we can obtain the radiance for each exit point. Then the object is rendered in a standard way with linear interpolation between the exit points.

8.4.1   Implementation

The light path map algorithm consists of a preprocessing step, which builds the LPM, and a rendering step, which takes advantage of the LPM to evaluate indirect illumination. The

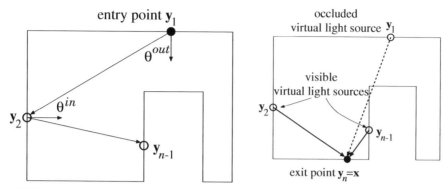

CPU: virtual light source generation.   GPU: illumination of the virtual lights.

FIGURE 8.8: Notations used in the formal discussion. The CPU is responsible for building a random path starting at the entry point. The visited points of the random path are considered as virtual point light sources. The GPU computes the illumination of virtual lights at exit points.

preprocessing step is implemented as a combined CPU ray-tracing and GPU method. On the other hand, the rendering step is realized completely on the GPU.

First, a CPU program samples entry points, and builds random paths starting from them. Uniformly distributed entry points can be generated on a triangle mesh by the following method. Random number r_1 that is uniformly distributed in $[0, 1]$ is obtained calling the pseudo-random generator of the CPU or taking the next member of a low-discrepancy series. This random value is multiplied by the sum of triangle areas $S = \sum_{i=1}^{N} A_i$, where A_i is the area of the triangle i and N is the number of triangles. Product $r_1 S$ is the *selection value* of the sampling. Triangle areas are summed until the running sum gets larger than the selection value, i.e. when

$$\sum_{i=1}^{j-1} A_i < r_1 S \leq \sum_{i=1}^{j} A_i.$$

When this happens, the jth triangle will be the selected triangle.

In the second phase of sampling, a random point is generated on the selected triangle. Suppose that the triangle vertices are $\mathbf{v}_1$, $\mathbf{v}_2$, and $\mathbf{v}_3$. Two new random values r_2 and r_3 are obtained from a uniform distribution in $[0, 1]$. The random entry point inside the triangle can be obtained by combining the edge vectors with these random values

$$\mathbf{e} = \mathbf{v}_1 + r_2(\mathbf{v}_2 - \mathbf{v}_1) + r_3(\mathbf{v}_3 - \mathbf{v}_1) \quad \text{if } r_2 + r_3 \leq 1.$$

If the sum of random numbers r_2, r_3 were greater than 1, they are replaced by $1 - r_2$ and $1 - r_3$, respectively.

The sampling process discussed so far generates entry point $\mathbf{e}$ with uniform distribution, i.e. its probability density is $p(\mathbf{e}) = 1/S$ where S is the total area of all surfaces in the scene. Having generated entry point $\mathbf{e}$, a random light path is started from here. The first step is a random decision with probability s_1 whether or not the path is declared to be terminated already at its start point. If the path is not terminated, a random direction is generated. The entry point and the random direction define a ray. The ray is traced and the hit point is identified. At the hit point we decide again randomly whether or not the light path is terminated with probability s_2. If the path needs to be continued, we repeat the same operation to find a new direction and trace the obtained ray. Sooner or later the random decisions will tell us to terminate the path, which thus has a random length (the method of random path termination is called *Russian roulette*). Since we are interested in the reflected radiance of the exit points, the points of the random path are connected to all exit points, which means that we get a number of paths ending in each exit point that is visible from the visited points of the path.

Let us denote the points of a path by $\mathbf{y}_1 = \mathbf{e}$ (entry point), $\mathbf{y}_2, \ldots, \mathbf{y}_{n-1}$ (randomly visited points), and $\mathbf{y}_n = \mathbf{x}$ (exit point) (Figure 8.8). Assuming unit irradiance at the entry point this path causes

$$f_1 G_1 \ldots f_{n-1} G_{n-1} f_n,$$

radiance at the exit point, where

$$G_k = G(\mathbf{y}_k, \mathbf{y}_{k+1}) = \frac{\cos \theta_{\mathbf{y}_k}^{\text{out}} \cdot \cos \theta_{\mathbf{y}_{k+1}}^{\text{in}}}{|\mathbf{y}_k - \mathbf{y}_{k+1}|^2} v(\mathbf{y}_k, \mathbf{y}_{k+1})$$

is the *geometry factor*. In the formula of the geometry factor, $\theta_{\mathbf{y}_k}^{\text{out}}$ is the angle between the surface normal at $\mathbf{y}_k$ and outgoing direction $\mathbf{y}_k \to \mathbf{y}_{k+1}$, $\theta_{\mathbf{y}_{k+1}}^{\text{in}}$ is the angle between the surface normal at $\mathbf{y}_{k+1}$ and incoming direction $\mathbf{y}_{k+1} \to \mathbf{y}_k$, *visibility function* $v(\mathbf{y}_k, \mathbf{y}_{k+1})$ indicates if the two points are not occluded from each other, and f_k is the diffuse BRDF of point $\mathbf{y}_k$.

In order to calculate the radiance at an exit point, all light paths ending here should be considered and their contribution added, which leads to a series of high-dimensional integrals adding light paths of different lengths. For example, the contribution of light paths of length n is

$$L_n^{\text{out}}(\mathbf{x}) = \int_S \cdots \int_S f_1 G_1 \cdots f_{n-1} G_{n-1} f_n \, d\mathbf{y}_1 \cdots d\mathbf{y}_{n-1}.$$

The Monte Carlo quadrature approximates these integrals by the average of the contributions of sample paths divided by the probability of their generation. Taking a single sample, the

Monte Carlo estimate is

$$L^{\text{out}} r_n(\mathbf{x}) \approx \frac{f_1 G_1 \dots f_{n-1} G_{n-1} f_n}{p(\mathbf{y}_1, \dots, \mathbf{y}_n)} = \frac{1}{p(\mathbf{y}_1)} \frac{f_1}{s_1} \frac{G_1}{p(\mathbf{y}_2|\mathbf{y}_1)} \frac{f_2}{s_2} \dots \frac{G_{n-2}}{p(\mathbf{y}_{n-1}|\mathbf{y}_{n-2})} f_{n-1} G_{n-1} f_n,$$

where the probability density of the path

$$p(\mathbf{y}_1, \dots, \mathbf{y}_n) = p(\mathbf{y}_1) s_1 p(\mathbf{y}_2|\mathbf{y}_1) s_2 \dots p(\mathbf{y}_{n-1}|\mathbf{y}_{n-2})$$

is expressed as the product of the probability density of the entry point $p(\mathbf{y}_1)$, the continuation probabilities s_k, and the conditional probability densities $p(\mathbf{y}_{k+1}|\mathbf{y}_k)$ of sampling $\mathbf{y}_{k+1}$ given previous sample $\mathbf{y}_k$. Note that $s_{n-1} p(\mathbf{y}_n|\mathbf{y}_{n-1})$ is not included in the probability since the exit point is connected to the random visited point deterministically.

According to the concept of *importance sampling*, we should sample the directions and continue the path to make the probability proportional to the integrand as much as possible, i.e. to reduce the variance of the random estimate. This means that the next point of the path should be sampled proportionally to the geometry factor, that is $p(\mathbf{y}_{k+1}|\mathbf{y}_k) \propto G(\mathbf{y}_k, \mathbf{y}_{k+1})$, and the path should be continued proportionally to the BRDF. Since the BRDF is not a scalar but depends on the wavelength, we make the continuation probability proportional to the *luminance* of the BRDF.

The next visited point $\mathbf{y}_{k+1}$ can be sampled proportionally to the geometry factor in the following way. First, the direction of a ray of origin $\mathbf{y}_k$ is sampled from cosine distribution of angle θ from the surface normal. Let us assign a coordinate system to current point $\mathbf{y}_k$ so that axis z is parallel with the surface normal, and axes x and y are in the plane of the triangle. The unit vectors of this coordinate system in world space are denoted by $\mathbf{T}$ (*tangent*), $\mathbf{B}$ (*binormal*), and $\mathbf{N}$ (*normal*). Let us now consider a unit radius sphere centered at $\mathbf{y}_k$. The sphere intersects the xy plane in a circle. Projecting a differential area on the sphere of size $d\omega$ onto the xy plane the area will be multiplied by $\cos\theta$ where θ is the angle between the sphere normal at $d\omega$ and the direction of the projection (axis z). This angle is equal to the angle between the direction into the center of $d\omega$ and axis z. If we sample the points in a unit radius circle uniformly, the probability of selecting a small area dA at x, y will be dA/π since π is the area of the unit radius circle. This means that we select the projection of solid angle $d\omega$ with probability $\cos\theta d\omega/\pi$. This is exactly what we need. The center of $d\omega$ is at unit distance from the center of the sphere and has x, y coordinates in the tangent plane, thus its z coordinate is $\sqrt{1 - x^2 - y^2}$. The sampled direction in world space is then

$$\omega = x\mathbf{T} + y\mathbf{B} + \sqrt{1 - x^2 - y^2}\mathbf{N}.$$

In order to generate uniformly distributed points in a unit radius circle, let us first sample two, independent values x and y in $[-1, 1]$, and check whether the point of these coordinates are inside the unit radius circle by examining $x^2 + y^2 \leq 1$. If not, this random pair is rejected, and another pair is sampled until we get a point that is really in the circle. The accepted point will have uniform distribution in the unit radius circle.

Having sampled the direction with this technique, the ray will leave the current point $\mathbf{y}_k$ in solid angle $d\omega$ with probability

$$P(d\omega) = \frac{\cos \theta_{\mathbf{y}_k}^{\text{out}}}{\pi} d\omega.$$

In solid angle $d\omega$ the surface of size

$$dy = d\omega \frac{|\mathbf{y}_k - \mathbf{y}_{k+1}|^2}{\cos \theta_{\mathbf{y}_{k+1}}^{\text{in}}}$$

is visible. Thus the probability density of hitting point $\mathbf{y}_{k+1}$ by the ray from $\mathbf{y}_k$ is

$$p(\mathbf{y}_{k+1}|\mathbf{y}_k) = \frac{P(d\omega)}{dy} = \frac{\cos \theta_{\mathbf{y}_k}^{\text{out}} \cos \theta_{\mathbf{y}_{k+1}}^{\text{in}}}{\pi |\mathbf{y}_k - \mathbf{y}_{k+1}|^2} = \frac{G(\mathbf{y}_k, \mathbf{y}_{k+1})}{\pi}.$$

For diffuse surfaces the albedo is π times the BRDF, i.e. $a_i = f_i \pi$. For physically plausible materials the albedo is less than 1 at any wavelength. Let us denote the luminance of the albedo by $\mathcal{L}(a_i)$ and set the continuation probability to be equal to this

$$s_i = \mathcal{L}(a_i) = \mathcal{L}(f_i)\pi.$$

Using cosine distributed direction sampling and taking the luminance of the albedo as the continuation probability, the random estimator of a single sample is

$$L_n^{\text{out}}(\mathbf{x}) \approx \frac{1}{p(\mathbf{y}_1)} \frac{a_1}{s_1} \frac{G_1}{p(\mathbf{y}_2|\mathbf{y}_1)} \cdots \frac{G_{n-2}}{p(\mathbf{y}_{n-1}|\mathbf{y}_{n-2})} f_{n-1} G_{n-1} f_n =$$

$$S \frac{f_1}{\mathcal{L}(f_1)} \cdots \frac{f_{n-2}}{\mathcal{L}(f_{n-2})} f_{n-1} G_{n-1} f_n.$$

The random path is generated by the CPU program, but the last step, the connection of the visited points and the exit points is done on the GPU. Note that a visited point can be imagined as a *virtual spot light source*. In direction ω enclosing angle $\theta_{\mathbf{y}_{n-1}}^{\text{out}}$ with the surface normal, this spot light source has intensity

$$S \frac{f_1}{\mathcal{L}(f_1)} \frac{f_2}{\mathcal{L}(f_2)} \cdots \frac{f_{n-2}}{\mathcal{L}(f_{n-2})} f_{n-1} \cos \theta_{\mathbf{y}_{n-1}}^{\text{out}}.$$

Thus, the GPU just computes the reflection of this intensity at exit points $\mathbf{x}$ that are visible from the virtual spot light source. This requires shadow mapping that can be efficiently implemented

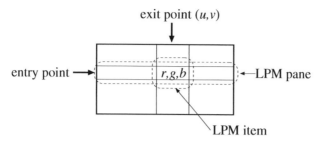

FIGURE 8.9: Representation of an LPM as an array indexed by entry points and exit points. A single element of this map is the LPM item, a single row is the LPM pane.

by the GPU. Since the exit points are the texels of a texture map, the virtual light source algorithm should render into a texture map. LPM items are computed by rendering into the texture with the random walk nodes as light sources and averaging the results of the random samples.

LPMs are stored in textures for real-time rendering. A single texel stores an LPM item that represents the contribution of all paths connecting the same entry point and exit point. An LPM can thus be imagined as an array indexed by entry points and exit points, and storing the radiance on the wavelengths of red, green, and blue (Figure 8.9). Since an exit point itself is identified by two texture coordinates (u, v), an LPM can be stored either in a 3D texture or in a set of 2D textures (Figure 8.10), where each one represents a single LPM pane (i.e. a row of the table in Figure 8.9, which includes the LPM items belonging to a single entry point).

The number of 2D textures is equal to the number of entry points. However, the graphics hardware has just a few texture units. Fortunately, this can be sidestepped by tiling the LPM panes into one or more larger textures.

Using the method as described above allows us to render indirect illumination interactively with a typical number of 256 entry points. While this figure is generally considered sufficient for a medium complexity scene, difficult geometries and animation may emphasize virtual light

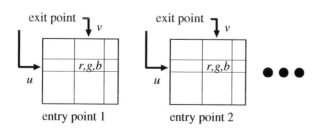

FIGURE 8.10: LPM stored as 2D textures.

source artifacts as spikes or flickering, thus requiring even more samples. Simply increasing the number of entry points and adding corresponding LPM panes would quickly challenge even the latest hardware in terms of texture memory and texture access performance. To cope with this problem, we can apply an approximation (a kind of lossy compression scheme), which keeps the number of panes under control when the number of entry points increase.

The key recognition is that if entry points are close and lay on similarly aligned surfaces, then their direct illumination will be probably very similar during the light animation, and therefore the LPMs of this *cluster* of entry points can be replaced by a single *average* LPM. Of course this is not true when a hard shadow boundary separates the two entry points, but due to the fact that a single entry point is responsible just for a small fraction of the indirect illumination, these approximation errors can be tolerated and do not cause noticeable artifacts. This property can also be understood if we examine how clustering affects the represented indirect illumination. Clustering entry points corresponds to a low-pass filtering of the indirect illumination, which is usually already low-frequency by itself, thus the filtering does not cause significant error. Furthermore, errors in the low-frequency domain are not disturbing for the human eye. Clustering also helps to eliminate animation artifacts. When a small light source moves, the illumination of an entry point may change abruptly, possibly causing flickering. If multiple entry points are clustered together, their average illumination will change smoothly. This way clustering also trades high-frequency error in the temporal domain for low-frequency error in the spatial domain.

To reduce the number of panes, contributions of a cluster of nearby entry points are added and stored in a single LPM pane. As these *clustered entry points* cannot be separated during rendering, they will all share the same weight when the entry point contributions are combined. This common weight is obtained as the average of the individual weights of the entry points. Clusters of entry points can be identified by the K-means algorithm [73] or, most effectively, by a simple object median splitting kd-tree. It is notable that increasing the number of samples via increasing the cluster size has only a negligible overhead during rendering, namely the computation of more weighting factors. The expensive access and combination of LPM items

FIGURE 8.11: A few tiles of an LPM texture used to render Figure 8.15. Illumination corresponding to clusters of entry points is stored in tiled atlases.

FIGURE 8.12: Entry points generated randomly.

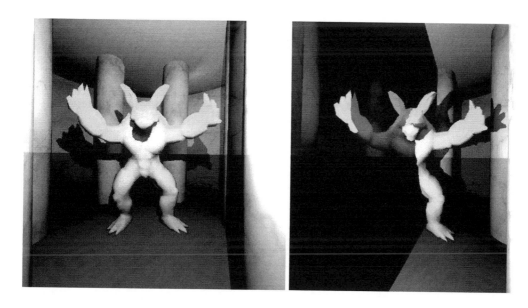

FIGURE 8.13: Comparison of local illumination and the light path map method. The lower half of these images has been rendered with local illumination, while the upper half with the light path map.

FIGURE 8.14: The chairs scene lit by a rectangular spot light. The rest is indirect illumination obtained with the light path map method at 35 FPS on an NV6800GT and close to 300 FPS on an NV8800.

is not affected. This way the method can be scaled up to problems of arbitrary complexity at the cost of longer preprocessing only.

While rendering the final image, the values stored in the LPM should be weighted according to the current lighting and be summed. Computing the weighting factors involves a visibility check that could be done using ray casting, but, as rendering direct illumination shadows would require a shadow map anyway, it can effectively be done in a shader, rendering to a one-dimensional texture of weights. Although these values would later be accessible

FIGURE 8.15: Escher staircases scenes lit by moving lights.

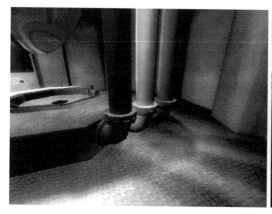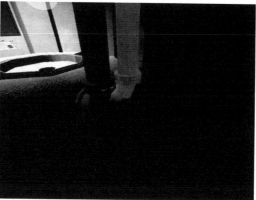

FIGURE 8.16: Indirect illumination computed by the light path map method in the Space Station game (left) and the same scene rendered with local illumination (right) for comparison. Note the beautiful *color bleeding* effects produced by the light path map.

via texture reads, they can be read back and uploaded into constant registers for efficiency. Furthermore, zero weight textures can be excluded, sparing superfluous texture accesses.

In order to find the indirect illumination at an exit point, the corresponding LPM items should be read from the textures and their values summed having multiplied them by the weighting factors and the light intensity. We can limit the number of panes to those having the highest weights. Selection of the currently most significant texture panes can be done on the CPU before uploading the weighting factors as constants.

Figures 8.12 and 8.13 show a marble chamber test scene consisting of 3335 triangles, rendered on 1024×768 resolution. We used 4096 entry points. Entry points were organized into 256 clusters. We set the LPM pane resolution to 256×256, and used the 32 highest weighted entry clusters. In this case, the peak texture memory requirement was 128 Mbytes. The constant parameters of the implementation were chosen to fit easily with hardware capabilities, most notably the maximum texture size and the number of temporary registers for optimized texture queries, but these limitations can be sidestepped easily. As shown in Figure 8.12, a high entry point density was achieved. Assuming an average albedo of 0.66 and starting two paths from each entry point, the 4096 entry points displayed in Figure 8.12 translate to approximately 24 000 virtual light sources.

For this scene, the preprocessing took 8.5 s, which can further be decomposed as building the kd-tree for ray casting (0.12 s), light tracing with CPU ray casting (0.17 s), and LPM generation (8.21 s). Having obtained the LPM, we could run the global illumination rendering interactively changing the camera and light positions. Figure 8.13 shows screen shots where one-half of the image was rendered with the light path map, and the other half with local illumination to allow comparisons. The effects are most obvious in shadows, but also note color bleeding and finer details in indirect illumination that could not be achieved by fake methods like using an ambient lighting term.

The role of the light path map method in interactive applications is similar to that of the *light map*. It renders indirect lighting of static geometry, but allows for dynamic lighting. The final result will be a plausible rendering of indirect illumination. Indirect shadows and color bleeding effects will appear, illumination will change as the light sources move (Figures 8.14, 8.15, and 8.16).

CHAPTER 9

Participating Media Rendering

So far we assumed that the radiance is constant along a ray and scattering may occur just on object surfaces. Participating media, however, may scatter light not only on their boundary, but anywhere inside their volume. Such phenomena are often called *subsurface scattering*, or *translucency*. Participating media can be imagined as some material that does not completely fill the space. Thus, the photons have the chance to go into it and to travel a random distance before collision. To describe light–volume interaction, the basic rendering equation should be extended. The volumetric rendering equation is obtained considering how the light goes through participating media (Figure 9.1).

The change of radiance L on a path of differential length ds and of direction ω depends on different phenomena:

Absorption: the light is absorbed when photons collide with the material and the material does not reflect the photon after collision. This effect is proportional to the number of photons entering the path, i.e. the radiance, the probability of collision, and the probability of absorption given that collision happened. If the probability of collision in a unit distance is τ, then the probability of collision along distance ds is τds. After collision a particle is reflected with the probability of *albedo a*, and absorbed with probability $1 - a$. Collision density τ and the albedo may also depend on the wavelength of the light. Summarizing, the total radiance change due to absorption is $-\tau ds \cdot (1 - a)L$.

Out-scattering: the light is scattered out from its path when photons collide with the material and are reflected after collision. This effect is proportional to the number of photons entering the path, the probability of collision, and the probability of reflection. The total out-scattering term is $-\tau ds \cdot aL$.

Emission: the radiance may be increased by the photons emitted by the participating media (e.g. fire). This increase is proportional to the density of the material, i.e. to τds, to the darkness of the material $(1 - a)$ since black bodies emit light more intensively than gray or white bodies, and to the emission density of an equivalent *black body radiator*, L^e. The total contribution of emission is $\tau ds \cdot (1 - a)L^e$.

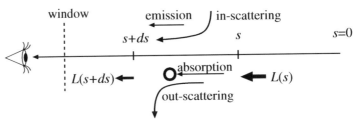

FIGURE 9.1: Modification of the radiance of a ray in participating media.

In-scattering: photons originally flying in a different direction may be scattered into the considered direction. The expected number of scattered photons from differential solid angle $d\omega'$ equals the product of the number of incoming photons and the probability that the photon is scattered from $d\omega'$ to ω in distance ds. The scattering probability is the product of the collision probability (τds), the probability of not absorbing the photon (a), and the probability density of the reflection direction ω, given that the photon arrived from direction ω', which is called *phase function* $P(\omega', \omega)$. Taking into account all incoming directions Ω', the radiance increase due to in-scattering is

$$\tau ds \cdot a \cdot \left(\int_{\Omega'} L^{in}(s, \omega') P(\omega', \omega) \, d\omega' \right).$$

Adding the discussed changes, we obtain the following *volumetric rendering equation* for radiance L of the ray at $s + ds$:

$$L(s + ds, \omega) = (1 - \tau ds) L(s, \omega) + \tau ds (1 - a) L^e(s, \omega)$$

$$+ \tau ds \cdot a \cdot \int_{\Omega'} L^{in}(s, \omega') P(\omega', \omega) \, d\omega'. \qquad (9.1)$$

Subtracting $L(s)$ from both sides and dividing the equation by ds, the volumetric rendering equation becomes an integro-differential equation:

$$\frac{dL(s, \omega)}{ds} = -\tau L(s, \omega) + \tau(1 - a) L^e(s, \omega) + \tau a \int_{\Omega'} L^{in}(s, \omega') P(\omega', \omega) \, d\omega'.$$

In homogeneous media, volume properties τ, a, and $P(\omega', \omega)$ are constant. In inhomogeneous media these properties depend on the actual position.

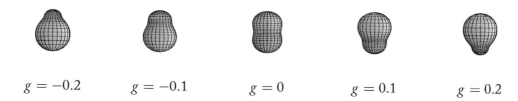

$$g = -0.2 \qquad g = -0.1 \qquad g = 0 \qquad g = 0.1 \qquad g = 0.2$$

FIGURE 9.2: The lobes of Henyey–Greenstein phase function for different g values. The light direction points upward.

9.1   PHASE FUNCTIONS

In the case of *diffuse scattering* the reflected radiance is uniform, thus the phase function is constant:

$$P_{\text{diffuse}}(\omega', \omega) = \frac{1}{4\pi}.$$

Note that constant $1/4\pi$ guarantees that the phase function is indeed a probability density, that is, its integral for all possible input (or output) directions equals 1.

Non-diffuse media reflect more photons close to the original direction (*forward scattering*) and backward where the light came from (*backward scattering*). A popular model to describe non-diffuse scattering is the *Henyey–Greenstein phase function* [65, 27]:

$$P_{HG}(\omega', \omega) = \frac{1}{4\pi} \frac{3(1 - g^2)(1 + (\omega'\omega)^2)}{2(2 + g^2)(1 + g^2 - 2g(\omega'\omega))^{3/2}}, \qquad (9.2)$$

where $g \in (-1, 1)$ is a material property describing how strongly the material scatters forward or backward (Figure 9.2).

9.2   PARTICLE SYSTEM MODEL

The *particle system model* of the volume corresponds to a discretization, when we assume that scattering can happen only at N discrete points called *particles*. Particles are sampled randomly, preferably from a distribution proportional to collision density τ, but may also be selected from a deterministic pattern. Particle systems can generate natural phenomena models of acceptable visual quality with a few hundred particles.

Let us assume that particle p represents its spherical neighborhood of diameter $\Delta s^{(p)}$, where the participating medium is approximately homogeneous, that is, its parameters are constant. This means that the volume around a particle center can be described by a few parameters. Let us first introduce the *opacity* of a particle as the probability that the photon collides when it goes through the sphere of this particle. The opacity is denoted by α and is computed as

$$\alpha^{(p)} = 1 - e^{-\int_{\Delta s^{(p)}} \tau(s) ds} = 1 - e^{-\tau^{(p)} \Delta s^{(p)}} \approx \tau^{(p)} \Delta s^{(p)}. \qquad (9.3)$$

The *emission* of a particle is its total emission accumulated in the particle sphere:

$$E^{(p)} = \int_{\Delta s^{(p)}} \tau(s)(1 - a(s))L^e(s) \, ds \approx \tau^{(p)} \Delta s^{(p)}(1 - a^{(p)})L^{e,(p)} \approx \alpha^{(p)}(1 - a^{(p)})L^{e,(p)}.$$

The average *incoming radiance* arriving at the sphere from a given direction is denoted by $I^{(p)}$. Similarly, the average *outgoing radiance* in a single direction is $L^{(p)}$. Using these notations, the *discretized volumetric rendering equation* at particle p is

$$L^{(p)}(\omega) = (1 - \alpha^{(p)})I^{(p)}(\omega) + E^{(p)}(\omega) + \alpha^{(p)}a^{(p)} \int_{\Omega'} I^{(p)}(\omega')P^{(p)}(\omega', \omega) \, d\omega'. \qquad (9.4)$$

The first term of this sum is the incoming radiance reduced by absorption and out-scattering, the second term is the emission, and the third term represents the in-scattering.

In homogeneous media opacity α, albedo a, and phase function P are the same for all particles. In inhomogeneous media, these parameters are particle attributes [117].

9.3   BILLBOARD RENDERING

Having approximated the in-scattering term somehow, the volume can efficiently be rendered from the camera by *splatting* particles on the screen and composing their contribution with *alpha blending*. To splat a single particle, we render a semi transparent, textured rectangle placed at the particle position and turned toward to the viewing direction. Such a rectangle is called a *billboard*. The texture of this rectangle should represent smoothing, so it is usually follows a Gaussian function. The in-scattering term is attenuated according to the total opacity of the particles that are between the camera and this particle. This requires particles be sorted in the view direction before sending them to the frame buffer in back to front order. At a given particle, the evolving image is decreased according to the opacity of the particle and increased by its in-scattering and emission terms.

Billboards are planar rectangles having no extension along one dimension. This can cause artifacts when billboards intersect opaque objects making the intersection of the billboard plane and the object clearly noticeable (Figure 9.3). The core of this *clipping artifact* is that a billboard fades those objects that are behind it according to its transparency as if the object were fully behind the sphere of the particle. However, those objects that are in front of the billboard plane are not faded at all, thus transparency changes abruptly at the object–billboard intersection.

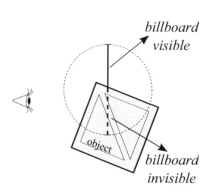

FIGURE 9.3: Billboard clipping artifact. When the billboard rectangle intersects an opaque object, transparency becomes spatially discontinuous.

On the other hand, when the camera moves into the participating media, billboards also cause *popping artifacts*. In this case, the billboard is either behind or in front of the front clipping plane, and the transition between the two stages is instantaneous. The former case corresponds to a fully visible, while the latter to a fully invisible particle, which results in an abrupt change during animation (Figure 9.4).

In the following section, we discuss a simple but effective solution for placing objects into the participating medium without billboard clipping and popping artifacts [149].

9.3.1 Spherical Billboards

Billboard clipping artifacts are solved by calculating the real path length a light ray travels inside a given particle since this length determines the opacity value to be used during rendering. The

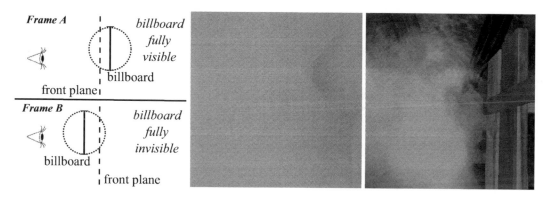

FIGURE 9.4: Billboard popping artifact. Where the billboard crosses the front clipping plane, the transparency is discontinuous in time (the figure shows two adjacent frames in an animation).

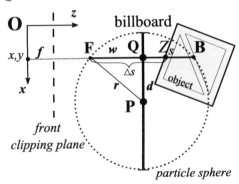

FIGURE 9.5: Computation of length Δs the ray segment travels inside a particle sphere in camera space.

traveled distance is obtained from the spherical geometry of particles instead of assuming that a particle can be represented by a planar rectangle. However, in order to keep the implementation simple and fast, we still send the particles through the rendering pipeline as quadrilateral primitives, and take their spherical shape into account only during fragment processing.

To find out where opaque objects are during particle rendering, first these objects are drawn and the resulting depth buffer storing camera space z coordinates is saved in a texture.

Having rendered the opaque objects, particle systems are processed. The particles are rendered as quads perpendicular to axis z of the camera coordinate system. These quads are placed at the farthest point of the particle sphere to avoid unwanted front plane clipping. Disabling depth test is also needed to eliminate incorrect object–billboard clipping.

When rendering a fragment of the particle, we compute the interval the ray travels inside the particle sphere in camera space. This interval is obtained considering the saved depth values of opaque objects and the camera's front clipping plane distance. From the resulting interval we can compute the opacity for each fragment in such a way that both fully and partially visible particles are displayed correctly, giving the illusion of a volumetric medium. During opacity computation we first assume that the density is uniform inside a particle sphere.

For the formal discussion of the processing of a single particle, let us use the notations of Figure 9.5. We consider the fragment processing of a particle of center $\mathbf{P} = (x_p, y_p, z_p)$, which is rendered as a quad perpendicular to axis z. Suppose that the current fragment corresponds to the visibility ray cast through point $\mathbf{Q} = (x_q, y_q, z_q)$ of the quadrilateral. Although visibility rays start from the origin of the camera coordinate system and thus form a perspective bundle, for the sake of simplicity, we consider them as being parallel with axis z. This approximation is acceptable if the perspective distortion is not too strong. If the visibility ray is parallel with axis z, then the z coordinates of $\mathbf{P}$ and $\mathbf{Q}$ are identical, i.e. $z_q = z_p$.

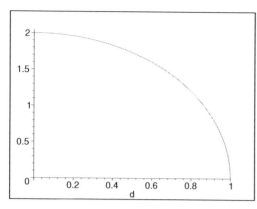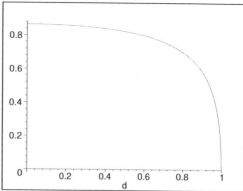

FIGURE 9.6: The accumulated density of a ray (left) and its seen opacity (right) as the function of the distance of the ray and the center in a unit sphere with constant, unit density.

The radius of the particle sphere is denoted by r. The distance between the ray and the particle center is $d = \sqrt{(x - x_p)^2 + (y - y_p)^2}$. The closest and the farthest points of the particle sphere on the ray from the camera are $\mathbf{F}$ and $\mathbf{B}$, respectively. The distances of these points from the camera can be obtained as

$$|\mathbf{F}| \approx z_p - w, \qquad |\mathbf{B}| \approx z_p + w,$$

where $w = \sqrt{r^2 - d^2}$. The ray travels inside the particle in interval $[|\mathbf{F}|, |\mathbf{B}|]$.

Taking into account the front clipping plane and the depth values of opaque objects, these distances may be modified. First, to eliminate popping artifacts, we should ensure that $|\mathbf{F}|$ is not smaller than the front clipping plane distance f, thus the distance the ray travels in the particle before reaching the front plane is not included. Second, we should also ensure that $|\mathbf{B}|$ is not greater than Z_s which is the stored object depth at the given pixel, thus the distance traveled inside the object is not considered.

From these modified distances, we can obtain the real length the ray travels in the particle:

$$\Delta s = \min(Z_s, |\mathbf{B}|) - \max(f, |\mathbf{F}|).$$

9.3.1.1 Gouraud Shading of Particles

We assumed that the density is homogeneous inside a particle and used equation (9.3) to obtain the respective opacity value. These correspond to piecewise constant finite-element approximation. While constant finite-elements might be acceptable from the point of view of numerical precision, their application results in annoying visual artifacts.

Figure 9.6 depicts the accumulated density ($\int_{\Delta s_j} \tau(s) ds$) and the respective opacity as a function of ray–particle distance d, assuming constant finite elements. Note that at the contour of the particle sphere ($d = 1$) the accumulated density and the opacity become zero, but they

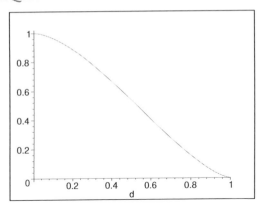

FIGURE 9.7: The accumulated density of a ray as the function of the distance of the ray and the center in a unit sphere assuming that the density function linearly decreases with the distance from the particle center.

do not diminish smoothly. The accumulated density has a significant derivative at this point, which makes the contour of the particle sphere clearly visible for the human observer.

This artifact can be eliminated if we use piecewise linear rather than piecewise constant finite-elements, that is, the density is supposed to be linearly decreasing with the distance from the particle center. The accumulated density computed with this assumption is shown by Figure 9.7. Note that this function is very close to a linear function, thus the effects of piecewise linear finite elements can be approximated by modulating the accumulated density by this linear function. This means that instead of equation (9.3) we use the following formula to obtain the opacity of particle j:

$$\alpha_j \approx 1 - e^{-\tau_j(1-d/r_j)\Delta s_j},$$

where d is the distance between the ray and the particle center, and r_j is the radius of this particle. Note that this approach is very similar to the trick applied in radiosity methods. While computing the patch radiosities piecewise constant finite elements are used, but the final image is presented with *Gouraud shading*, which corresponds to linear filtering.

9.3.1.2 GPU Implementation of Spherical Billboards

The evaluation of the length the ray travels in a particle sphere and the computation of the corresponding opacity can be efficiently executed by a custom fragment shader program. The fragment program gets some of its inputs from the vertex shader: the particle position in camera space (P), the shaded billboard point in camera space (Q), the particle radius (r), and the shaded point in the texture of the depth values (depthuv). The fragment program also gets uniform parameters, including the texture of the depth values of opaque objects (Depth), the density

FIGURE 9.8: Particle system rendered with planar (left) and with spherical (right) billboards.

(tau), and the camera's front clipping plane distance (f). The fragment processor calls the following function to obtain the opacity of the given particle at this fragment:

```
float SphericalBillboard( in float3 P,       // Particle center
                          in float3 Q,       // Shaded billboard point
                          in float2 depthuv, // Texture coordinate in the depth texture
                          in float f,        // Front clipping plane
                          in float r,        // Particle radius
                          in float tau ) {   // Density
    float alpha = 0;
    float d = length(P.xy - Q.xy);           // Distance from the particle center
    if(d < In.r){                            // If the ray crosses the particle sphere
        float zp = P.z;
        float w  = sqrt(r*r - d*d);          // Half length of ray segment in the sphere
        float F  = zp - w;                   // Ray enters the sphere here
        float B  = zp + w;                   // Ray exits the sphere here
        Zs = tex2D(Depth, depthuv).r;        // Depth of opaque objects
        float ds = min(Zs, B) - max(f, F);   // Corrected traveling length
        alpha = 1 - exp(-tau * (1-d/r) * ds);  // Opacity with Gouraud shading
    }
    return alpha;
}
```

With this simple calculation, the shader program obtains the real ray segment length (ds) and computes opacity alpha of the given particle that controls blending of the particle into the frame buffer. The consideration of the spherical geometry during fragment processing eliminates clipping and popping artifacts (see Figure 9.8).

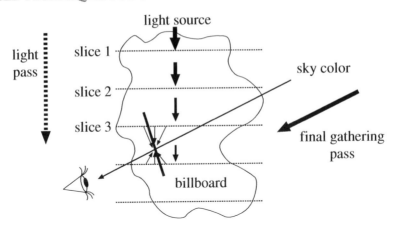

FIGURE 9.9: Final gathering for a block.

9.4   ILLUMINATING PARTICIPATING MEDIA

Illuminating participating media consists of a separate *light pass* for each light source determining the approximation of the in-scattering term caused by this particular light source, and a *final gathering* step. Handling light–volume interaction on the particle level would be too computation intensive since the light pass requires the incremental rendering of all particles and the read back of the actual result for each of them. To speed up the process, we render particle blocks one by one, and separate light–volume interaction calculation from the particles [148].

During a light pass we sort particles according to their distances from the light source, and organize particles of similar distances in particle groups. Particles are rendered according to the distance, and the evolving image is saved into a texture after completing each group. These textures are called *slices* [56]. The first texture will display the accumulated opacity of the first group of particles, the second will show the opacity of the first and second groups of particle blocks, and so on. Figure 9.9 shows this technique with five depth slices. Four slices can be computed simultaneously if we store the slices in the color channels of one RGBA texture. For a given particle, the vertex shader will decide in which slice (or color channel) this particle should be rendered. The fragment shader sets the color channels corresponding to other slices to zero, and the pixel is updated with alpha blending.

During final rendering, we sort particles and render them one after the other in back to front order. The in-scattering term is obtained from the sampled textures of the slices that enclose the pixel of the block (Figure 9.9).

The accumulated opacity of the slices can be used to determine the radiance at the pixels of these slices and finally the reflected radiance of a particle between the slices. Knowing the position of the particles we can decide which two slices enclose it. By linear interpolation

FIGURE 9.10: Comparison of participating media rendering with spherical billboards and the discussed illumination method (left) and with the classic single scattering model (right) in the Space Station game. Note that the classic method does not attenuate light correctly and exhibits billboard clipping artifacts.

between the values read from these two textures we can approximate the attenuation of the light source color. The radiance of those pixels of both enclosing slices are taken into account, for which the phase function is not negligible.

Figure 9.10 shows the result of the discussed participating media rendering algorithm and compares it to the classic approach that has no light or shadowing pass. Note that the classic approach does not attenuate the light before scattering, thus the result will be unrealistically bright. Images of game scenes rendered by this method are shown in Figures 12.6 and 12.13.

Note that the discussed volume shader produces *volume shadows* and a limited form of *multiple scattering*, but cannot take into account all types of light paths. It simulates mostly forward scattering except in the vicinity of the view ray.

9.5   RENDERING EXPLOSIONS AND FIRE

In this section, we discuss the rendering of realistic *explosion* [149]. An explosion consists of *dust*, *smoke*, and *fire*, which are modeled by specific particle systems. Dust and smoke absorb light, fire emits light. These particle systems are rendered separately, and the final result is obtained by compositing their rendered images.

9.5.1   Dust and Smoke

Smoke particles are responsible for absorbing light in the fire, and typically have low albedo values ($a = 0.2$, $\tau = 0.4$). High albedo ($a = 0.9$, $\tau = 0.4$) dust swirling in the air, on the other hand, is added to improve the realism of the explosion (Figure 9.11).

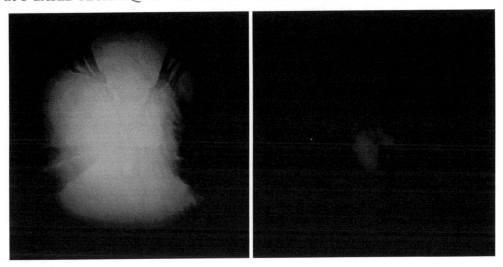

FIGURE 9.11: High albedo dust and low albedo smoke.

When rendering dust and smoke we assume that these particles do not emit radiance so their emission term is zero. To calculate the in-scattering term, the length the light travels in the particle sphere, the albedo, the density, and the phase function are needed. We can use the Henyey–Greenstein phase function (equation (9.2)). To speed up the rendering, these function values are fetched from a prepared 2D texture addressed by $\cos\theta = \omega' \cdot \omega$ and g, respectively. Setting g to constant zero gives satisfactory results for dust and smoke. The real length the light travels inside a smoke or dust particle is computed by the spherical billboard method.

In order to maintain high frame rates, the number of particles should be limited, which may compromise high detail features. To cope with this problem, the number of particles is reduced while their radius is increased. The variety needed by the details is added by perturbing the opacity values computed by spherical billboards. Each particle has a unique, time-dependent perturbation pattern. The perturbation is extracted from a grey scale texture, called *detail image*, which depicts real smoke or dust (Figure 9.12). The perturbation pattern of a particle is taken from a randomly placed, small quad shaped part of this texture. This technique has been used for long by off-line renderers of the motion picture industry [4]. As time advances this texture is dynamically updated to provide variety in the time domain as well. Such animated 2D textures can be obtained from real world videos and stored as a 3D texture since inter-frame interpolation and looping can automatically be provided by the graphics hardware's texture sampling unit [96].

9.5.2   Fire

Fire is modeled as a black body radiator rather than participating medium, i.e. its albedo is zero, so only the emission term is needed. The color characteristics of fire particles are determined

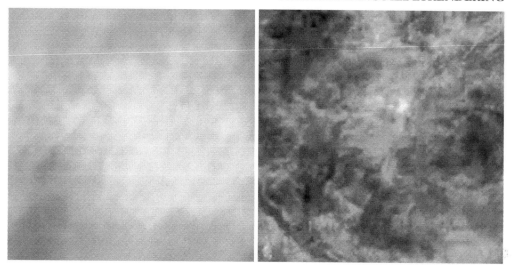

FIGURE 9.12: Images from real smoke and fire video clips, which are used to perturb the billboard fragment opacities and temperatures.

by the physics theory of *black body radiation*. For wavelength λ, the emitted radiance of a black body can be computed by *Planck's formula*:

$$L_\lambda^e(x) = \frac{2C_1}{\lambda^5(e^{C_2/(\lambda T)} - 1)},$$

where $C_1 = 3.7418 \times 10^{-16}$ W m^2, $C_2 = 1.4388 \times 10^{-2}$ m K, and T is the absolute temperature of the radiator [126]. Figure 9.13 shows the spectral radiance of black body radiators at different temperatures. Note that the higher the temperature is, the more blueish the color gets.

For different temperature values, the RGB components can be obtained by integrating the spectrum multiplied by the *color matching functions*. These integrals can be precomputed and stored in a texture. To depict realistic fire, the temperature range of $T \in [2500\ \text{K}, 3200\ \text{K}]$ needs to be processed (see Figure 9.14).

High detail features are added to fire particles similarly to smoke and dust particles. However, now not the opacity, but the emission radiance should be perturbed. We could use a color video and take the color samples directly from this video, but this approach would limit the freedom of controlling the temperature range and color of different explosions. Instead, we can store the temperature variations in the detail texture (Figure 9.12), and its stored temperature values are scaled and are used for color computation on the fly. A randomly selected, small quadrilateral part of a frame in the detail video is assigned to a fire particle to control the

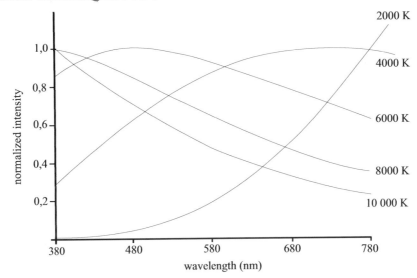

FIGURE 9.13: Black body radiator spectral distribution.

temperature perturbation of the fragments of the particle billboard. The temperature is scaled and a bias is added if required. Then the resulting temperature is used to find the color of this fragment in the black body radiation function.

The fragment program gets fire particle position in camera space (P), the shaded billboard point in camera space (Q), the particle radius (r), the location of the shaded point in the depth texture (depthuv), and the position of the detail image and the starting time in the video in detail. The fragment program also gets uniform parameters, including the texture of the fire video (FireVideo), the black body radiation function of Figure 9.14 (BBRad), the depth values of opaque objects (Depth), the density (tau), the camera's front clipping plane distance (f), and the temperature scale and bias in T1 and T0, respectively.

The fragment processor computes the color of the fire particle with the following HLSL program that uses spherical billboards discussed in section 9.3.1.2:

FIGURE 9.14: Black body radiator colors from 0 K to 10,000 K. Fire particles belong to temperature values from 2500 K to 3200 K.

```
float4 FireFP( in float3 P       : TEXCOORD0, // Particle position in camera space,
               in float3 Q       : TEXCOORD1, // Shaded billboard point in camera space
               in float1 r       : TEXCOORD2, // Particle radius
               in float2 depthuv : TEXCOORD4, // Location in the depth texture
               uniform float time,            // Animation time
               uniform sampler3D FireVideo,   // Detail video
               uniform sampler1D BBRad,       // Black body radiator functions
               uniform sampler2D Depth,       // Depth map
               uniform float tau,             // Fire density
               uniform float f,               // Front clipping plane distance
               uniform float T0,              // Temperature bias
               uniform float T1               // Temperature scale
             ) : COLOR0
{
    float alpha = SphericalBillboard(P, Q, depthuv, f, r, tau);
    float3 detuvw = detail + float3(0,0,time);      // Address for the detail video
    float T = T0 + T1 * tex3D(FireVideo, detuvw);   // Temperature
    return float4(tex1D(BBRad, T).rgb, 1) * alpha;  // Color
}
```

9.5.3   Layer Composition

To combine the particle systems together and with the image of opaque objects, a layer composition method can be used. This way we should render the opaque objects and the particle systems into separate textures, and then compose them. This leads to three rendering passes: one pass for opaque objects, one pass for dust, fire, and smoke, and one final pass for composition. The first pass computes both the color and the depth of opaque objects.

One great advantage of rendering the participating medium into a texture is that we can use floating point blending. Another advantage is that this render pass may have a smaller resolution render target than the final display resolution, which considerably speeds up rendering.

To enhance realism, we can also simulate *heat shimmering* that distorts the image [96]. This is done by confusing the image according to a noisy texture. This noise is used in the final composition as *u*, *v* offset values to distort the image. With the help of multiple render targets, this pass can also be merged into the pass of rendering of fire particles.

The final effect that could be used through composition is *motion blur*, which can easily be done with blending, letting the new frame fade into previous frames. The complete rendering process is shown in Figure 9.15. The dust, the fire, and the smoke consist of 16, 115, and 28 animated particles, respectively. Note that these surprisingly small number of larger particles can be simulated very efficiently, but thanks to the opacity and temperature perturbation, the high-frequency details are not compromised. The modeled scene consists of 16 800 triangles. The scene is rendered with per pixel Phong shading. The algorithm offers real-time rendering speed

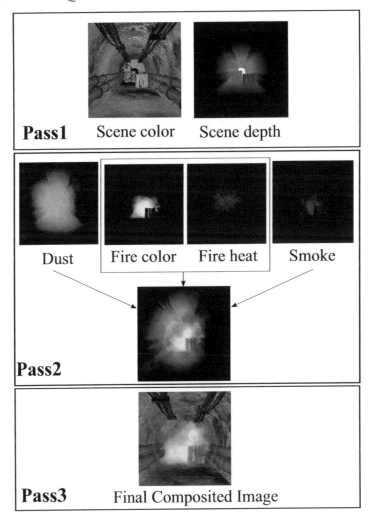

FIGURE 9.15: Explosion rendering algorithm.

(40 FPS) (see Figure 9.16). For comparison, the scene without the particle system is rendered at 70 FPS, and the classic billboard rendering method would also run at about 40 FPS. This means that the performance lost due to the more complex spherical billboard calculations can be regained by decreasing the render target resolution during particle system drawing.

9.6   PARTICIPATING MEDIA ILLUMINATION NETWORKS

This section discusses a real-time method to compute multiple scattering in non-homogeneous participating media having general phase functions. The volume represented by a particle system is supposed to be static, but the lights and the camera may move. Real-time performance is achieved by reusing light scattering paths that are generated with global line bundles traced

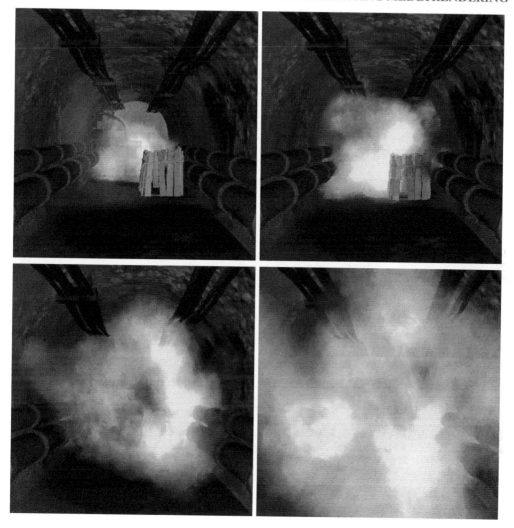

FIGURE 9.16: Rendered frames from an explosion animation sequence.

in sample directions in a preprocessing phase [142]. For each particle, we obtain those other particles which can be seen in one of the sample directions, and their radiances toward the given particle. This information allows the fast iteration of the volumetric rendering equation.

9.6.1 Iteration Solution of the Volumetric Rendering Equation

The global illumination in participating media is mathematically equivalent to the solution of the volumetric rendering equation. If the participating medium is represented by particles, the integro-differential form of the volumetric rendering equation can be presented in a simpler approximate form (equation (9.4)).

The method discussed in this section solves the discretized volumetric rendering equation by iteration. Note that unknown particle radiance values appear both on the left and the right sides of this equation. If we have an estimate of particle radiance values (and consequently, of incoming radiance values), then these values can be inserted into the formula of the right side and a new estimate of the particle radiance can be provided. Iteration keeps repeating this step. If the process is convergent, then in the limiting case the formula would not alter the radiance values, which are therefore the roots of the equation. Formally, the new radiance of particle p ($p = 1, \ldots, N$) in iteration step m is obtained by substituting the radiance values of iteration step $m - 1$ to the right side of the discretized volumetric rendering equation:

$$L_m^{(p)}(\omega) = (1 - \alpha^{(p)})I_{m-1}^{(p)}(\omega) + E^{(p)}(\omega) + \alpha^{(p)}a^{(p)} \int_{\Omega'} I_{m-1}^{(p)}(\omega')P^{(p)}(\omega', \omega)\, d\omega'. \qquad (9.5)$$

The convergence of this iteration is guaranteed if the opacity is in $[0, 1]$ and the albedo is less than 1, which is always the case for physically plausible materials.

In order to calculate the directional integral representing the in-scattering term of equation (9.5), we suppose that D random directions $\omega_1', \ldots, \omega_D'$ are obtained from uniform distribution of density $1/(4\pi)$, and the integral is estimated by the Monte Carlo quadrature:

$$\int_{\Omega'} I^{(p)}(\omega')P^{(p)}(\omega', \omega)\, d\omega' \approx \frac{4\pi}{D} \sum_{d=1}^{D} I^{(p)}(\omega_d')P^{(p)}(\omega_d', \omega).$$

9.6.2   Building the Illumination Network

If we use the same set of sample directions for all particles, then the incoming radiance and therefore the outgoing radiance are needed only at these directions during iteration. For a single particle p, we need D incoming radiance values $I^{(p)}$ in $\omega_1', \ldots, \omega_D'$, and the reflected radiance needs to be computed exactly in these directions. In order to update the radiance of a particle, we should know the indices of the particles visible in sample directions, and also the distances of these particles to compute the opacity (Figures 9.17 and 9.18). This information can be stored in two-dimensional arrays $\mathbf{I}$ and $\mathbf{V}$ of size $N \times D$, indexed by particles and directions, respectively (Figure 9.19). Array $\mathbf{I}$ is called the *illumination network* and stores the incoming radiance values of the particles on the wavelengths of red, green, and blue. Array $\mathbf{V}$ is the *visibility network* and stores index of visible particle vp and opacity α for each particle and incoming direction, that is, it identifies from where the given particle can receive illumination.

In order to handle emissions and the direct illumination of light sources, we can use a third array $\mathbf{E}$ that stores the sum of the emission and the reflection of the direct illumination for each particle and discrete direction. This array can be initialized by rendering the volume from the point of view of the light source and identifying those particles that are directly visible. At

FIGURE 9.17: Two directions of the visibility network.

a particular particle, the reflection of the light source is computed for each outgoing discrete direction.

Visibility network **V** expressing the visibility between particles is constructed during a preprocessing phase. First random directions are sampled. For each direction, particle centers are orthographically projected onto the window that is perpendicular to the direction. Discretizing the window, the particles that are projected onto a particular pixel are identified and are sorted according to their distance from the window. Particles that follow each other in the sorted list form pairs of particles that occlude each other in the given direction. On the other hand, the difference of the depth values is the distance of the particles, from which the opacity can be computed.

9.6.3 Iterating the Illumination Network

The solution of the global illumination problem requires the iteration of the illumination network. A single step of the iteration evaluates the following formula for each particle

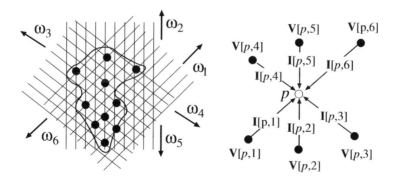

FIGURE 9.18: A single particle in the illumination and visibility networks.

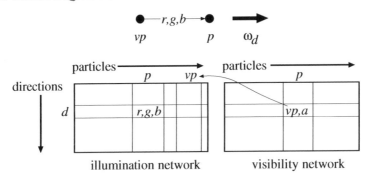

FIGURE 9.19: Storing the networks in arrays.

$p = 1, \ldots, N$ and for each incoming direction $i = 1, \ldots, D$:

$$\mathbf{I}[p, i] = (1 - \alpha_{\mathbf{V}[p,i]})\mathbf{I}[\mathbf{V}[p, i], i] + \mathbf{E}[\mathbf{V}[p, i], i]$$

$$+ \frac{4\pi\, \alpha_{\mathbf{V}[p,i]} a_{\mathbf{V}[p,i]}}{D} \sum_{d=1}^{D} \mathbf{I}[\mathbf{V}[p, i], d]\, P_{\mathbf{V}[p,i]}(\omega'_d, \omega_i).$$

Interpreting the two-dimensional arrays of the emission, visibility, and illumination maps as textures, the graphics hardware can also be exploited to update the illumination network. We work with three textures, the first two are static and the third is updated in each iteration. The first texture is visibility network $\mathbf{V}$ storing the visible particle in the red and the opacity in the green channels, the second stores emission array $\mathbf{E}$ in the red, green, and blue channels, and the third texture is the illumination network, which also has red, green, and blue channels. Note that in practical cases the number of particles N is about a thousand, while the number of sample directions D is typically 128, and radiance values are half precision floating point numbers, thus the total size of these textures is quite small ($1024 \times 128 \times 8 \times 2$ bytes $= 2$ Mbyte).

In the GPU implementation a single iteration step is the rendering of a viewport-sized, textured rectangle, having set the viewpoint resolution to $N \times D$ and the render target to the texture map representing the updated illumination network. A pixel corresponds to a single particle and single direction, which are also identified by input variable `texcoord`. The fragment shader obtains the visibility network from texture `Vmap`, the emission array from texture `Emap`, and the illumination map from texture `Imap`. Function P is responsible for the phase function evaluation, which is implemented by a texture lookup of prepared values [117]. In this simple implementation we assume that opacity `alpha` is pre-computed and stored in the visibility texture, but albedo `alb` are constant for all particles. Should it not be the case, the albedo could also be looked up in a texture.

When no other particle is seen in the input direction, then the incoming illumination is taken from the sky radiance (sky). In this way not only a single sky color, but sky illumination textures can also be used [107, 117].

For particle p and direction i, the fragment shader finds opacity alpha and visible particle vp in direction i, takes its emission or direct illumination Evp, and computes incoming radiance Ip as the sum of the direct illumination and the reflected radiance values for its input directions d = 1, . . . ,D (Figure 9.20):

```
float4 IllumNetworkFP(
                in float2 texcoord  : TEXCOORD0, // Particle p and input direction i
                uniform float3 sky,              // Sky illumination
                uniform sampler2D Vmap,          // Visibility map
                uniform sampler2D Emap,          // Emission map
                uniform sampler2D Imap           // Illumination map
                ) : COLOR0
{
    float  p = texcoord.x;      // Particle
    float  i = texcoord.y;      // Input direction
    float  vp = tex2d(Vmap, float2(p,i)).r; // Visible particle or zero
    float3 Ip;
    if (vp >= 0) {                      // If vp exists
        float alpha = tex2d(Vmap, float2(p,i)).g;   // Opacity between this and vp
        float3 Evp = tex2d(Emap, float2(vp,i)).rgb; // Emission of vp
        float3 Ivp = tex2d(Imap, float2(vp,i));     // Radiance from the same direction
        float3 Ip = (1 - alpha) * Ivp + Evp;        // Transmitted radiance of vp
        for(int d = 0; d < 1; d += 1.0/D) {         // Reflected radiance of vp
            Ivp = tex2d(Imap, float2(vp, d));
            float3 BRDF = alb * alpha * P(d,i);
            Ip +=  BRDF * Ivp * 4 * PI / D;
        }
    } else Ip = sky;            // Incident radiance is the sky illumination
    return float4(Ip, 1);       // Incident radiance radiance of p
}
```

The illumination network provides a view-independent radiance representation. When the final image is needed, we can use a traditional participating media rendering method, which sorts the particles according to their distance from the camera, splats them, and adds their contributions with alpha blending.

When the outgoing reflected radiance of a particle is needed, we compute the reflection from the sampled incoming directions to the viewing direction. Finally, the sum of particle emission and direct illumination of the external lights is interpolated from the sample directions, and is added to the reflected radiance.

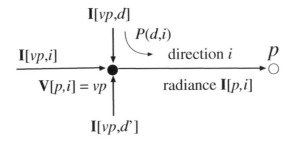

FIGURE 9.20: Notations in the fragment shader code.

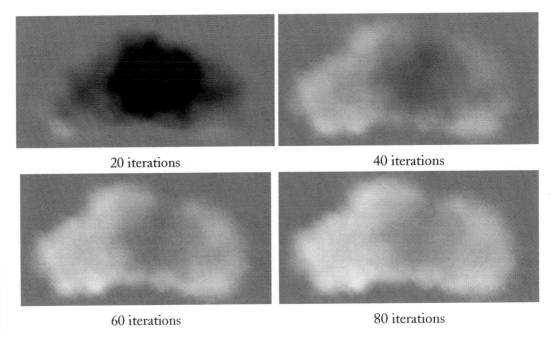

20 iterations

40 iterations

60 iterations

80 iterations

FIGURE 9.21: A cloud illuminated by two directional lights rendered with different iteration steps.

FIGURE 9.22: Globally illuminated clouds of 512 particles rendered with 128 directions.

The results are shown in Figures 9.21 and 9.22. The cloud model consists of 1024 particles, and 128 discrete directions are sampled. With these settings the typical rendering speed is about 26 frames per second, and is almost independent of the number of light sources and of the existence of sky illumination. The albedo is 0.9 and material parameter g is 0. In Figure 9.21, we can follow the evolution of the image of the same cloud after different iteration steps, where we can observe the speed of convergence.

CHAPTER 10

Fake Global Illumination

Global illumination computation is inherently costly since all other points may affect the illumination of a particular point. The inherent complexity can be reduced if the physically plausible model is replaced by another model that is simpler to solve but provides somehow similarly looking results. The dependence of the radiance of a point on all other points can be eliminated if we recognize that the illumination influence of a unit area surface diminishes with the square of the distance (refer to the geometry factor in equation (1.9)), thus it is worth considering only the local neighborhood of each point during shading and replacing the rest by some average. A point is said to be "open" in directions where the local neighborhood is not visible from it. In open directions, a constant or direction-dependent *ambient* illumination is assumed, which represents the "rest". In not open directions, the ambient light is assumed to be blocked.

Instead of working with the physically based reflected radiance formula of equation (1.7), fake global illumination methods approximate the indirect illumination according to the level of openness of points. The direct illumination can be obtained by local illumination algorithms, and added to the approximated indirect illumination solution.

Several related methods were born that are based on this idea. The *obscurances method* [168, 67]—in its original, scalar form—computes just how "open" the scene is in the neighborhood of a point, and scales the ambient light accordingly. The method called *ambient occlusion* [62, 104, 81] also approximates the solid angle and the average direction where the neighborhood is open, thus we can use environment map illumination instead of the ambient light. In the *spectral obscurances method* the average spectral reflectivities of the neighborhood and the whole scene are also taken into account, thus even *color bleeding* effects can be cheaply simulated [92]. Recent work [18] extended ambient occlusion as well to incorporate color bleeding.

Here we discuss only the original scalar obscurances method and the spectral obscurances approach.

10.1   THE SCALAR OBSCURANCES METHOD

The scalar obscurances method approximates the indirect illumination as

$$L^{\text{ind}}(\mathbf{x}) = a(\mathbf{x})W(\mathbf{x})L^a, \tag{10.1}$$

where $a(\mathbf{x})$ is the *albedo* of the point, L^a is the ambient light intensity, and $W(\mathbf{x})$ is a scalar factor in $[0, 1]$ that expresses how open point $\mathbf{x}$ is. Recall that for diffuse surfaces, the albedo is π times the BRDF (equation (1.12)).

Scalar factor $W(\mathbf{x})$ is called the *obscurance* of point $\mathbf{x}$, and defined as

$$W(\mathbf{x}) = \frac{1}{\pi} \int_{\omega' \in \Omega'} \rho(d(\mathbf{x}, \omega')) \cos^+ \theta'_{\mathbf{x}} \, d\omega', \tag{10.2}$$

where $\rho(d)$ is the scaling of the ambient light incoming from distance d and $\theta'_{\mathbf{x}}$ is the angle between direction ω' and the normal at $\mathbf{x}$. Function $\rho(d)$ increases with distance d allowing the ambient light to take effect if no occluder can be found nearby. Let us introduce the maximum distance $d_{\max}$ to define what "nearby" means. We take into account only a $d_{\max}$-neighborhood of $\mathbf{x}$. In other words, we are not taking into account occlusions farther than $d_{\max}$. Intersections closer than $d_{\max}$, on the other hand, reduce the effect of ambient lighting. Ambient occlusion uses a step function for this, which is 0 if the distance is smaller than $d_{\max}$ and 1 otherwise. We can also apply some gradual change using, for example, the function

$$\rho(d) = \sqrt{d/d_{\max}} \quad \text{if } d < d_{\max} \text{ and 1 otherwise.}$$

Since $0 \le \rho(d) \le 1$, the resulting obscurance value will be between 0 and 1. An obscurance value of 1 means that the point is totally open (or not occluded by neighboring polygons), while a value of 0 means that it is totally closed (or occluded by neighboring polygons).

For a closed environment, the ambient light in equation (10.1) can be approximated in the following way. The average emission and the albedo in the scene are

$$\tilde{L}^e = \frac{\sum_{i=1}^n A_i L_i^e}{S}, \quad \tilde{a} = \frac{\sum_{i=1}^n A_i a_i}{S},$$

where A_i, L_i^e, and a_i are the area, emission radiance, and the albedo of patch i, respectively, S is the sum of the areas, and n is the number of patches in the scene. The average radiance after a single reflection will roughly be $\tilde{a}\tilde{L}^e$. Similarly, the average radiance after two, three, etc. reflections are $\tilde{a}^2\tilde{L}^e$ and $\tilde{a}^3\tilde{L}^e$, etc. Adding up all reflections, we obtain an approximation of the ambient light:

$$L^a = \tilde{a}\tilde{L}^e + \tilde{a}^2\tilde{L}^e + \tilde{a}^3\tilde{L}^e + \cdots = \frac{\tilde{a}}{1 - \tilde{a}}\tilde{L}^e. \tag{10.3}$$

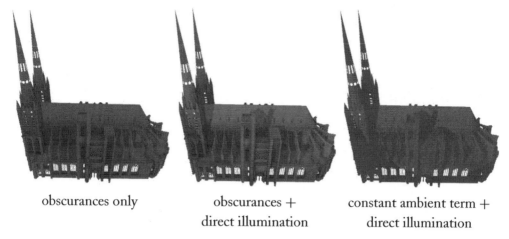

obscurances only   obscurances +   constant ambient term +
                      direct illumination   direct illumination

FIGURE 10.1: The cathedral rendered with the scalar obscurances algorithm and also by the standard ambient + direct illumination model (right) for comparison.

10.2 THE SPECTRAL OBSCURANCES METHOD

The obscurances approach presented so far lacks one of the global illumination features, namely *color bleeding*. Since the light reflected from a patch acquires some of its color, the surrounding patches receive colored indirect lighting. To account for color bleeding, the obscurances formula (equation (10.2)) is modified to include the diffuse BRDF of points visible from **x**:

$$W(\mathbf{x}) = \int_{\omega' \in \Omega'} f_r(\mathbf{y})\rho(d(\mathbf{x}, \omega')) \cos^+ \theta' \, d\omega', \qquad (10.4)$$

where $f_r(\mathbf{y})$ is the diffuse BRDF of point **y** that is visible from **x** in direction ω'. When no surface is seen at a distance less than d_{max} in direction ω', the spectral obscurance takes the value of average albedo $\tilde{a}$. For coherency, the ambient light expressed by equation (10.3) also has to be modified, yielding the following approximation:

$$L^a = \frac{1}{1 - \tilde{a}} \frac{\sum_{i=1}^{n} A_i L_i^e}{S}. \qquad (10.5)$$

In Figure 10.2, the Cornell box scene is shown computed by the obscurances method with and without color bleeding.

If the obscurances are available for surface points that correspond to texel centers of a texture map, called the *obscurances map*, then the application of the obscurances during rendering is similar to the application of *light maps*. The scene should be processed by a standard vertex

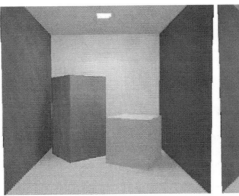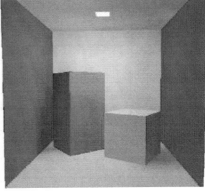

FIGURE 10.2: Comparison of spectral obscurances (right) to scalar obscurances (left). Note the color bleeding that can only be rendered with spectral obscurances.

shader. The fragment shader is responsible for direct illumination, possibly computing per-pixel lighting and shadow mapping, and for the addition of the indirect term according to the obscurances map.

```
sampler2D ObscuranceMap; // Texture map of the obscurances
sampler2D AlbedoMap;     // Texture map storing the albedo
float3 La;               // Ambient light intensity
float3 I;                // Directional or point light source intensity

float4 ObscurancePS(
      in float2 Tex  : TEXCOORD0, // Texture coordinates
      in float3 N    : TEXCOORD1, // Interpolated normal
      in float3 L    : TEXCOORD2  // Interpolated illumination direction
   ) : COLOR {
   // Compute per-pixel direct illumination with no ambient lighting
   float3 Color = ComputeDirectIllumination(Tex, AlbedoMap, N, L, I);
   // Add obscurances instead of the ambient light
   Color += La * tex2D(AlbedoMap, Tex) * tex2D(ObscuranceMap, Tex);
   return float4(Color, 1);
}
```

If there is just a single point light source in the scene, the realism can be further increased by weighting the spectral obscurance of a point with the distance from the light source (Figure 10.6).

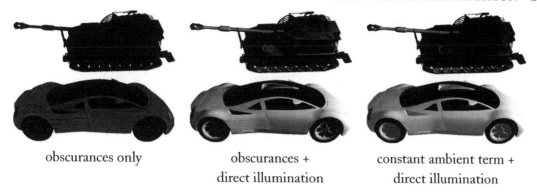

obscurances only   obscurances +   constant ambient term +
direct illumination   direct illumination

FIGURE 10.3: A tank and a car rendered with the spectral obscurances algorithm and also by the standard ambient + direct illumination model (right) for comparison.

10.3   CONSTRUCTION OF THE OBSCURANCES MAP

The obscurance computation is usually an off-line process, which generates a texture map, called the *obscurances map*. A texel of the texture map corresponds to a small *surface patch*. When a patch is referenced, we can simply use the texture address of the corresponding texel. A texel of the obscurances map stores the obscurance value of the point corresponding to the texel center, or the average obscurance of points of the small surface patch belonging to the texel.

The obscurances map construction can be implemented both on the CPU and on the GPU. The integral of equation (10.4) is usually estimated using the Monte Carlo quadrature. We cast several random rays from the examined point. The random direction is sampled from probability density $\cos^+ \theta'/\pi$ according to the concept of *importance sampling*. The obscurance estimate of the point will then be the average of the values gathered by the rays and divided by

FIGURE 10.4: Trees rendered with obscurances. In the left image the obscurances are applied to the left tree, but not to the right one to allow comparisons.

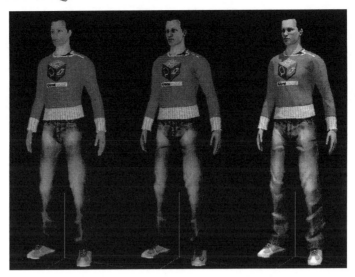

FIGURE 10.5: Digital Legend's character rendered with ambient illumination (left), obscurances map (middle), and obscurances and direct illumination (right).

the probability density of the sample:

$$W(\mathbf{x}) \approx \frac{1}{M} \sum_{j=1}^{M} \frac{f_r(\mathbf{y}_j)\rho(d_j)\cos^+\theta'_j}{\cos^+\theta'_j/\pi} = \frac{1}{M} \sum_{j=1}^{M} \rho(d_j)f_r(\mathbf{y}_j)\pi = \frac{1}{M} \sum_{j=1}^{M} \rho(d_j)a(\mathbf{y}_j),$$

(10.6)

where M is the number of rays, $\mathbf{y}_j$ is the point hit by ray j, and d_j is the length of this ray.

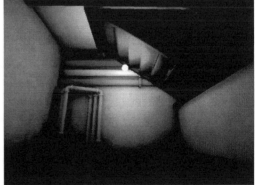

FIGURE 10.6: Two snapshots of a video animating a point light source, rendered by spectral obscurances. To increase the realism the obscurance of a point is weighted by the distance from the light sources.

10.4 DEPTH PEELING

Since casting cosine distributed rays from all points in the scene is equivalent to casting *global ray bundles* of uniformly distributed random directions [120], we can also cast global ray bundles (Figure 10.7). The term *global ray* means that we use all ray–surface intersections, and not only the hit point closest to the ray origin as in the case of *local rays*. A bundle includes many rays that are parallel with each other and their hit points would be uniformly distributed on a perpendicular plane.

Setting orthogonal projection and coloring each patch with their unique ID, we can suppose the pixel image to be equivalent to tracing a bundle of parallel rays through the scene where each pixel corresponds to a ray in the bundle. Each of these rays may intersect several surfaces in the scene. Normally, the depth buffer algorithm would identify the first intersections of these rays. However, rendering the scene several times and ignoring those points in later passes that have already been found, we can discover all the intersections in the form of image layers.

This multipass process is called *depth peeling* [39, 141, 52, 89]. Let us suppose that the scene consists of triangles that are further decomposed to small surface patches corresponding to the texels. For every pixel of the orthogonally projected image, the goal of depth peeling is to determine all of those texels where the corresponding surface patch is projected onto the pixel. In addition to the patch IDs we also need to know whether a patch is front-facing or back-facing with respect to the projection direction since during obscurance computation only those surfaces are taken into account that occlude the front face. We use the *pixel* of an image layer to store the patch ID, a flag indicating whether the patch is front-facing or back-facing to the camera, and the camera to patch distance.

The facing direction of a triangle can be determined by checking the sign of the z coordinate of its normal vector (note that this is the real normal vector of the rendered triangle and not the shading normal that is usually passed with vertices). Another alternative is to read the VFACE register of the fragment shader. A negative value indicates back-facing, a positive value front-facing triangle.

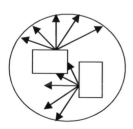
Rays from every patch

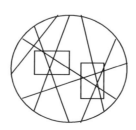
Global lines

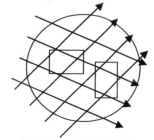
B undles of parallel rays

FIGURE 10.7: Different sampling techniques to generate cosine distributed rays.

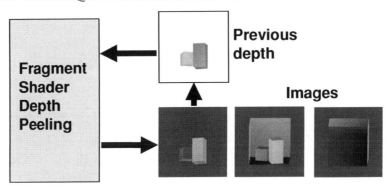

FIGURE 10.8: Depth peeling process.

As we store the pixels in a four-component float array, we can use the first two components to store the texture address of the patch, the third to store the facing information, and the fourth component to store the depth value.

The vertex shader receives the coordinates of a triangle vertex, the texture coordinates identifying the surface patch at this vertex, and generates the transformed vertex position both in clipping space and in the texture space of the previous layer. Transformation ModelViewProjTex is the concatenation of the model-view, projection, and the clipping space to texture space transformation $\mathbf{T}_{Tex}$ (equation (4.1)). When matrix $\mathbf{T}_{Tex}$ is constructed we should use a negative bias.

```
void DepthPeelVS(
    in  float4 Pos   : POSITION,   // Input position in modeling space
    in  float2 Tex   : TEXCOORD0,  // Patch ID as texture coordinates
    out float4 hPos  : POSITION,   // Output position in screen space
    out float2 oTex  : TEXCOORD0,  // Patch ID
    out float3 tPos  : TEXCOORD2   // Texture space + depth
    )
{
    hPos = mul(Pos, ModelViewProj);
    oTex = Tex;
    tPos = mul(Pos, ModelViewProjTex);  // Texture coordinates of the stored layer
}
```

The fragment shader receives the interpolated texture coordinates, which identify a small surface patch inside the processed triangle, and the interpolated depth of the patch. The peeling process is controlled by the depth value. The first layer stores those points that are visible in the pixel of the camera, and can be obtained by the depth buffer algorithm. Let us save the depth buffer of this step in a texture. The second layer can be explored by rendering the scene again

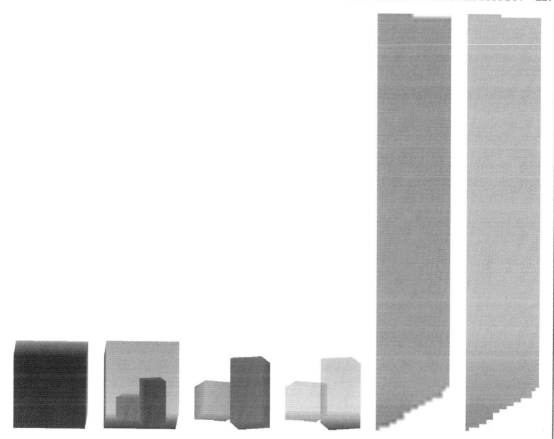

FIGURE 10.9: Six different image layers showing depth information for each pixel for the Cornell Box scene.

with the depth buffer algorithm, but discarding all fragments where the depth is (approximately) equal to the value stored in the depth buffer of the previous rendering step. Repeating the same step and discarding fragments that are not farther from the camera than the previous depth, we can generate the layers one by one. The fragment shader also stores the facing information obtained from register VFACE in the layer. This rendering step is repeated until all pixels are discarded.

```
uniform sampler2D prevLayer;    // Previous layer
uniform bool      first;        // Is first layer?

float4 DepthPeelPS(
    float  facing : VFACE,       // If facing < 0 then back face, otherwise front face
```

```
    float2 Tex    : TEXCOORD0, // Patch ID
    float3 tPos   : TEXCOORD2  // Point in texture space + depth
    ) : COLOR0
{
    if( !first ) { // not first -> peel
        float lastDepth = tex2D(prevLayer, tPos.xy).a;   // Last depth
        if (depth < lastDepth) discard;                  // Ignore previous layers
    }
    return float4(Tex.xy, facing, depth);                // Patch ID + facing + depth
}
```

A unique property of obscurances and ambient occlusion is that they can mimick global illumination transport both on the *macrostructure* level, where the geometry is defined as a triangle mesh, and on the *mesostructure level* where the geometry is represented by *bump maps*, *normal maps*, and *displacement maps* [143].

Thus, it is worth computing the scalar obscurances for a displacement map and use the result to modulate the macrostructure global illumination solution obtained with any other method.

CHAPTER 11

Postprocessing Effects

The result of a global illumination computation is the radiance seen through each pixel of the screen. Additionally, the depth of the visible points can also be fetched from the depth buffer. The radiance at a given wavelength is a floating point value, and at least three wavelengths are needed to render a color image. Radiance values may differ by many orders of magnitude, i.e. they have *high dynamic range* (*HDR*). However, the frame buffer containing the displayed image may store unsigned bytes, that can represent *low dynamic range* images (*LDR*). The conversion of HDR image values to displayable LDR values is called *tone mapping* [115].

During this conversion, we can also take into account that real world cameras and the human eye may cause lots of side-effects that contribute to the final appearance of a picture. Such effects include *depth of field*, *lens flare*, *glare*, *halo*, *coronae*, and *glow*. Postprocessing, in general, is responsible for tone mapping and for the generation of these camera effects. These methods require weighted averages of pixel values at different neighborhoods, which can be obtained by convolving the image with a particular filter kernel. Thus, the efficient implementation of filtering is a key to the fast realization of such effects. Filtering is implemented by rendering a *viewport-sized quad* to get the fragment shader to process every pixel.

In this chapter we first consider the efficient realization of 2D image filters, in particular the Gaussian filter. Then we present the implementations of postprocessing effects, including temporal tone mapping, glow, and depth of field.

11.1 IMAGE FILTERING

Image filtering is the most critical part of screen space methods. If we do it naively, the fragment shader needs to access the texture memory many times to fetch values in the neighborhood of the processed pixel. The general form of an image filter is

$$\tilde{L}(X, Y) = \int\limits_{y=-\infty}^{\infty} \int\limits_{x=-\infty}^{\infty} L(X - x, Y - y)w(x, y) \, \mathrm{d}x \, \mathrm{d}y,$$

where $\tilde{L}(X, Y)$ is the filtered value at pixel X, Y, $L(X, Y)$ is the original image, and $w(x, y)$ is the *filter kernel*. For example, the *Gaussian filter* of variance σ^2 has the following kernel:

$$w(x, y) = \frac{1}{2\pi\sigma^2} e^{-\frac{x^2+y^2}{2\sigma^2}}.$$

Since the exponential function diminishes quickly, the infinite integrals can be approximated by finite integrals in interval $[-S, S]$:

$$\tilde{L}(X, Y) \approx \int_{y=-S}^{S} \int_{x=-S}^{S} L(X - x, Y - y) w(x, y) \, dx dy.$$

An integral is replaced by a sum of N pixel samples for discrete images:

$$\tilde{L}(X, Y) \approx \sum_{i=-N/2}^{N/2-1} \sum_{j=-N/2}^{N/2-1} L(X - i, Y - j) w(i, j). \qquad (11.1)$$

This discrete integral approximation requires the evaluation of N^2 kernel values, multiplications, and additions, which is rather costly when repeated for every pixel of the screen.

In order to implement this naive approach, a single *viewport-sized quad* needs to be rendered, which covers all pixels of the screen. In clipping space, the viewport-sized quad is defined by vertices $(-1, -1, 0), (-1, 1, 0), (1, 1, 0), (1, -1, 0)$. Vertex shader ViewportSizedQuadVS generates the corresponding texture coordinates, taking care of that the whole texture is in a square of corners $(0, 0)$ and $(1, 1)$, the y-axis of texture space points downwards, and the center of a texel is a half texel far from its corner. The horizontal and vertical resolutions of the texture are represented by global variables HRES and VRES, respectively.

```
float HRES, VRES;

void ViewportSizedQuadVS( in  float4 Pos  : POSITION,   // Input corner of the quad
                          out float4 hPos : POSITION,   // Output in clipping space
                          out float2 oTex : TEXCOORD0 ) // Vertex in texture space
{
    hPos = Pos; // already in clipping space
    oTex.x = (1 + Pos.x)/2 + 0.5f/HRES;
    oTex.y = (1 - Pos.y)/2 + 0.5f/VRES;
}
```

Since a viewport-sized quad covers all pixels, it makes sure that the fragment shader is called for every pixel of the render target. The fragment shader is responsible for the evaluation of equation (11.1) for each pixel. The input image is texture InputImage. The current fragment is identified by the texture coordinates passed in variable Tex. The filter kernel w2 is realized according to the Gaussian filter.

```
sampler2D InputImage;  // Image to be filtered
float      sigma2;     // Variance of the Gaussian
int        N;          // Kernel width

const float ONE_PER_TWOPI = 1/2/3.1415;

float w2(in float2 xycoord) {  // 2D Gaussian filter
    return ONE_PER_TWOPI/sigma2 * exp(-dot(xycoord, xycoord)/2/sigma2);
}

float4 FilterPS(in float2 Tex : TEXCOORD0) : COLOR {
    float3 filtered = float3(0, 0, 0);
    for(int i = -N/2, i < N/2; i++) {
        for(int j = -N/2, j < N/2; j++) {
            float2 uv = float2(i/HRES, j/VRES);
            filtered += tex2D(InputImage, Tex - uv) * w2(uv);
        }
    }
    return float4(filtered, 1);
}
```

11.1.1 Separation of Dimensions

One common way of reducing the computation burden of 2D filtering is to exploit the *separability* of the filter kernel, which means that the two-variate filter kernel can be expressed in the following product form:

$$w(x, y) = w_1(x)w_1(y),$$

where w_1 is a single-variate function. The Gaussian filter is separable. The reduction of the computation for separable filters is based on the recognition that the two-dimensional convolution can be replaced by a vertical and a horizontal one-dimensional convolutions. The double integral is computed in two passes. The first pass results in the following 2D function:

$$L_h(X, Y) = \int_{-S}^{S} L(X - x, Y)w_1(x) \, dx.$$

Then the final result can be obtained by filtering L_h again with a similar one-dimensional filter:

$$\tilde{L}(X, Y) \approx \int_{-S}^{S} L_h(X, Y - y)w_1(y) \, dy.$$

In this way, the computation complexity can be reduced from N^2 to $2N$.

The implementation of the separable filter uses the same vertex shader (`ViewportSizedQuadVS`) as the naive implementation. However, the viewport-sized quad should be rendered two times, and two different fragment shaders should be activated in these passes. Fragment shader `FilterHPS` executes the horizontal filtering, while `FilterVPS` filters vertically.

```
sampler2D InputImage; // Image to be filtered
float     sigma;      // Standard deviation of the Gaussian
int       N;          // Kernel width

const float ONE_PER_SQRT_TWOPI = 0.3989;

float w1(in float x) {  // 1D Gaussian filter
    return ONE_PER_SQRT_TWOPI/sigma/sigma * exp(-x*x/2/sigma/sigma);
}

float4 FilterHPS(in float2 Tex0 : TEXCOORD0) : COLOR {
    float3 filtered = float3(0, 0, 0);
    for(int i = -N/2, i < N/2; i++) {
        float d = i/HRES;
        filtered += tex2D(InputImage, Tex - float2(d, 0)) * w1(d);
    }
    return float4(filtered, 1);
}

float4 FilterVPS(float2 Tex0 : TEXCOORD0) : COLOR {
    float3 filtered = float3(0, 0, 0);
    for(int i = -N/2, i < N/2; i++) {
        float d = i/VRES;
        filtered += tex2D(InputImage, Tex - float2(0, d)) * w1(d);
    }
    return float4(filtered, 1);
}
```

11.1.2 Exploitation of the Bi-linear Filtering Hardware

During convolution a weighted sum of texture values is computed. The most time consuming part of this computation is fetching the texture memory. To reduce the number of texture fetches, we can exploit the *bi-linear interpolation* feature of texture units, which takes four samples and computes a weighted sum requiring the cost of a single texture fetch [53].

As an example, let us consider a one-dimensional convolution using four samples

$$\tilde{L}(X) \approx \int_{x=-S}^{S} L(X-x)w(x)\,dx \approx$$

$$L(X-1)w(1) + L(X)w(0) + L(X+1)w(-1) + L(X+2)w(-2),$$

where L is the image (texture) value and w is the filter kernel, assuming that texels are addressed by integer coordinates. If the texture memory is fetched between two texel centers and bi-linear filtering is enabled, then we get a weighted average of the two texels. This means that instead of four texture memory fetches, the same result can be obtained with two fetches, where one fetch is between $X-1$ and X, and the other fetch is between $X+1$ and $X+2$. Let us denote the difference of these locations and X by h_0 and h_1, respectively (Figure 11.1). Turning bi-linear filtering on, the first fetch returns

$$F(X-1) = L(X-1)h_0 + L(X)(1-h_0).$$

The second fetch gives

$$F(X+1) = L(X+1)(2-h_1) + L(X+2)(h_1-1).$$

Values $F(X-1)$ and $F(X+1)$ are weighted and added to compute the desired filtered value $\tilde{L}(X)$. Let us denote the not yet known weights by g_{-1} and g_1, respectively. The weighted sum is then

$$\tilde{F}(X) = F(X-1)g_{-1} + F(X+1)g_1.$$

Fetch locations h_0 and h_1, as well as new weights g_{-1} and g_1 should be selected to make $\tilde{F}(X)$ equal to the desired filtered value $\tilde{L}(X)$. Inspecting the factors of $L(X-1)$, $L(X)$, $L(X+1)$,

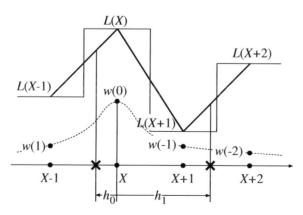

FIGURE 11.1: The idea of the exploitation of the bi-linear filtering hardware.

and $L(X+2)$ one by one, we get the following system of equations for unknown locations and weights:

$$
\begin{aligned}
g_{-1}h_0 &= w(1), \\
g_{-1}(1 - h_0) &= w(0), \\
g_1(2 - h_1) &= w(-1), \\
g_1(h_1 - 1) &= w(-2).
\end{aligned}
$$

Solving this equation, locations h_0, h_1 and weights g_{-1}, g_1 can be obtained. Note that this equation should be solved only once for each filter kernel, and the results can be stored in constants or in textures.

The presented idea halves the number of texture fetches from one-dimensional textures. In the case of two-dimensional textures, the number of texture fetches can be reduced to one-fourth.

11.1.3 Importance Sampling

In this section, we discuss another trick to further reduce the required computations [146]. This approach is based on the concept of *importance sampling*. Instead of sampling the integration domain regularly, importance sampling takes more samples where the filter kernel is large. Let us consider the $\int_{-S}^{S} L(X - x)w(x)\,dx$ convolution, and find integral $\tau(x)$ of the kernel and also its inverse $x(\tau)$ so that the following conditions hold:

$$
\frac{d\tau}{dx} = w(x) \;\rightarrow\; \tau(x) = \int_{-\infty}^{x} w(t)\,dt.
$$

If w is known, then τ can be computed and inverted off line. Substituting the precomputed $x(\tau)$ function into the integral we obtain

$$
\int_{-S}^{S} L(X - x)w(x)\,dx = \int_{\tau(-S)}^{\tau(S)} L(X - x(\tau))\,d\tau.
$$

Approximating the transformed integral taking uniformly distributed samples corresponds to a quadrature of the original integral taking N' non-uniform samples:

$$
\int_{\tau(-S)}^{\tau(S)} L(X - x(\tau))\,d\tau \approx \frac{|\tau(S) - \tau(-S)|}{N'} \sum_{i=1}^{N'} L(X - x(\tau_i)),
$$

where $|\tau(S) - \tau(-S)|$ is the size of the integration domain.

This way we take samples densely where the filter kernel is large and fetch samples less often farther away, but do not apply weighting. Note that this allows us to use a smaller number of samples ($N' < N$) and not to access every texel in the neighborhood since far from the center of the filter kernel, the weighting would eliminate the contribution anyway, so taking dense samples far from the center would be waste of time.

The implementation of this approach is quite straightforward. The required $x(\tau)$ function is computed by integrating the standard Gaussian function and inverting the integral. For the Gaussian filter, $\tau(x)$ is the cumulative probability distribution function of the normal distribution, i.e. the famous Φ-function,

$$\tau(x) = \Phi(x/\sigma).$$

Values of function $\Phi^{-1}(\tau)$ are hardwired into the shader for $\tau = 1/6, 2/6, 3/6, 4/6, 5/6$, respectively:

$$(-0.9924, -0.4307, 0, 0.4307, 0.9924).$$

These constants scaled with σ determine where the texture should be fetched from.

The convolution is executed separately for the two directions (this is possible because of the separability of the Gaussian filter). The horizontal and the vertical passes are implemented by the following fragment shaders:

```
sampler2D InputImage; // Image to be filtered
float sigma;          // Standard deviation of the Gaussian filter

float4 FilterImportanceHPS(in float2 Tex0 : TEXCOORD0) : COLOR {
    float2 du1 = float2(0.4307/HRES * sigma, 0);
    float2 du2 = float2(0.9924/HRES * sigma, 0);
    float3 filtered = tex2D(InputImage, Tex0 - du2) +
                      tex2D(InputImage, Tex0 - du1) +
                      tex2D(InputImage, Tex0) +
                      tex2D(InputImage, Tex0 + du1) +
                      tex2D(InputImage, Tex0 + du2);
    return float4(filtered/5, 1);
}
float4 FilterImportanceVPS(float2 Tex0 : TEXCOORD0) : COLOR {
    float2 dv1 = float2(0, 0.4307/VRES * sigma);
    float2 dv2 = float2(0, 0.9924/VRES * sigma);
    float3 filtered = tex2D(InputImage, Tex0 - dv2) +
                      tex2D(InputImage, Tex0 - dv1) +
                      tex2D(InputImage, Tex0) +
                      tex2D(InputImage, Tex0 + dv1) +
```

```
                    tex2D(InputImage, Tex0 + dv2);
     return float4(filtered/5, 1);
}
```

11.2   GLOW

Glow or *bloom* occurs when a very bright object in the picture causes the neighboring pixels to be brighter than they would be normally. It is caused by scattering in the lens and other parts of the eye, giving a glow around the light and dimming contrast elsewhere.

To produce glow, first we distinguish pixels where glowing parts are seen from the rest. After this pass we use Gaussian blur to distribute glow in the neighboring pixels, which is added to the original image (Figure 11.2). We can also add an interesting trailing effect to the glow. If we store the blurred glow image, we can modulate the glow image of the next frame with it. We can control the length of the trail during the composition with a dimming parameter.

11.3   TONE MAPPING

Off the shelf monitors can produce light intensity just in a limited, *low dynamic range* (*LDR*). Therefore the values written into the frame buffer are unsigned bytes in the range of [0x00, 0xff], representing values in [0,1], where 1 corresponds to the maximum intensity of the monitor. However, global illumination computations result in *high dynamic range* (*HDR*) luminance values that are not restricted to the range of the monitors. The conversion of HDR image values to displayable LDR values is called *tone mapping* [115]. The conversion is based on the *luminance* the human eye is adapted to. Assuming that our view spans over the image, the *adaptation luminance* will be the average luminance of the whole image. The luminance value of every pixel is obtained with the standard *CIE XYZ* transform (*D65 white point*):

$$Y = 0.2126R + 0.7152G + 0.0722B,$$

and these values are averaged to get adaptation luminance $\tilde{Y}$.

Having adaptation luminance $\tilde{Y}$, pixel luminance values Y are first mapped to *relative luminance* Y_r:

$$Y_r = \frac{\alpha Y}{\tilde{Y}}, \qquad (11.2)$$

where α is a constant of the mapping, which is called the *key value*.

original image without glow

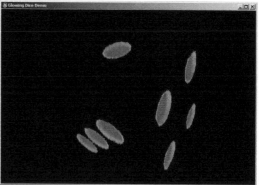

the glowing parts of the image

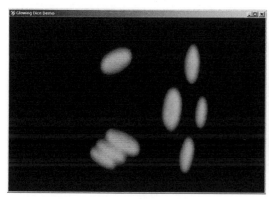

glow image with normal gain

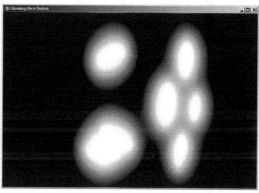

glow image with high gain

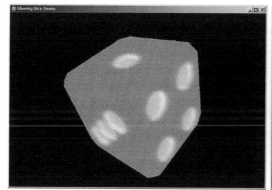

final image

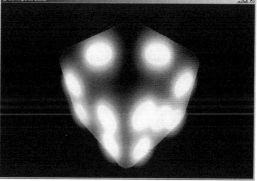

final image with extreme glow

FIGURE 11.2: The glow effect.

The relative luminance values are then mapped to displayable $[0,1]$ pixel intensities L using the following function:

$$L = \frac{Y_r}{1 + Y_r}.$$
(11.3)

This formula maps all luminance values to the $[0, 1]$ range in such way that relative luminance $Y_r = 1$ is mapped to pixel intensity $L = 0.5$. This property is used to map a desired luminance level of the scene to the middle intensity on the display by controlling the key value. Mapping a higher luminance level to middle gray results in a subjectively dark image, whereas mapping lower luminance to middle gray will give a bright result. Images which we perceive at low light condition are relatively dark compared to what we see during a day.

The exact key value α can be left as a user choice, or it can be estimated automatically based on the relations between minimum, maximum, and average luminance in the scene [113]. Unfortunately, the critical changes in the absolute luminance values may not always affect the relation between these three values. For example, this may lead to dark night scenes appearing as too bright.

Krawczyk [82] proposed an empirical method to calculate the key value. His method is based on the absolute luminance. Since the key value was introduced in photography, there is no scientifically based experimental data which would provide an appropriate relation between the key value and the luminance. The low key is 0.05, the typical choice for moderate illumination is 0.18, and 0.8 is the high key. Based on these, Krawczyk empirically specified the following formula:

$$\alpha(\tilde{Y}) = 1.03 - \frac{2}{2 + \log_{10}(\tilde{Y} + 1)}.$$
(11.4)

This basic tone mapping process can be extended in several ways. In the following subsections, we discuss a local version, the incorporation of the glow effect into the tone mapping process, the extension to temporally varying image sequence, and to cope with scotopic vision.

11.3.1 Local Tone Mapping

The tone mapping function of equation (11.3) may lead to the loss of details in the scene due to extensive contrast compression. Reinhard et al. [113] proposed a solution to preserve local details by employing a spatially variant local adaptation value V in equation (11.3):

$$L(x, y) = \frac{Y_r(x, y)}{1 + V(x, y)},$$

where x, y are the pixel coordinates.

Local adaptation V equals the average luminance in a neighborhood of the pixel. The problem lies however in the estimation of how large the neighborhood of the pixel should be. The goal is to have as wide neighborhood as possible, however too large area may lead to well-known inverse gradient artifacts called the *halo*. The solution is to successively increase the size of the neighborhood on each scale of the pyramid, checking each time if no artifacts are introduced. For this purpose, a Gaussian pyramid is constructed with successively increasing kernel size.

11.3.2 Glow Integration Into Tone Mapping

To take into account the additional light scattering of glow during the tone mapping process, we have to create a glow map based on the average luminance of the picture. The glow map selects those pixels whose luminance is significantly higher than the average. Having obtained the glow map it is blurred with a Gaussian filter to deliver illumination to the pixels close to the glowing parts.

In the final rendering step, we tone map the HDR image with the following equation:

$$L(x, y) = \frac{Y_r + Y_{\text{glow}}}{1 + V(x, y)},$$

where L is the final pixel intensity value, Y_r is the relative luminance, Y_{glow} is the value of the blurred glow map that represents the amount of additional light due to glow, and V is the local adaptation map.

11.3.3 Temporal Luminance Adaptation

While tone mapping a sequence of HDR frames, it is important to note that the luminance conditions can change drastically from frame to frame. The human vision system reacts to such changes through the *temporal adaptation*. The time course of the adaptation differs depending on whether we adapt to light or to darkness, and whether we perceive mainly using rods (during night) or cones (during a day).

To take into account the adaptation process, a filtered $\tilde{Y}_a$ value can be used instead of the actual adaptation luminance $\tilde{Y}$. The filtered value changes according to the adaptation processes in human vision, eventually reaching the actual value if the adaptation luminance is stable for some time. The process of adaptation can be modeled using an exponential decay function:

$$\tilde{Y}_a^{new} = \tilde{Y}_a + (\tilde{Y} - \tilde{Y}_a)(1 - e^{-\frac{T}{\tau}}),$$

where T is the discrete time step between the display of two frames, and τ is the time constant describing the speed of the adaptation process. These time constants are different for rods and

FIGURE 11.3: The temporal adaptation process in three frames of an animation.

cones:

$$\tau_{\text{rods}} \approx 0.4 \text{ s}, \qquad \tau_{\text{cones}} \approx 0.1 \text{ s}.$$

Therefore, the speed of the adaptation depends on the level of the illumination in the scene. The time required to reach the fully adapted state depends also on whether the observer is adapting to light or dark conditions. The above numbers describe the adaptation to light. The full adaptation to dark takes up to tens of minutes, so it is usually not simulated.

11.3.4 Scotopic Vision

In low light conditions only the rods are active, so color discrimination is not possible. The image becomes less colorful. The cones start to loose sensitivity at $3.4\frac{cd}{m^2}$ and become completely insensitive at $0.03\frac{cd}{m^2}$ where the rods are dominant. We can model the sensitivity of rods with the following equation:

$$\beta(Y) = \frac{\beta_0}{\beta_0 + Y}, \tag{11.5}$$

where Y denotes the luminance and β_0 is a user-defined constant that should depend on how bright our monitor is when $Y = 1$. The value $\beta(Y) = 1$ corresponds to the monochromatic vision and $\beta(Y) = 0$ the full color discrimination.

We account for the scotopic vision, while we scale the RGB channels according to the ratio of the displayed and original luminance values of the pixel. Using the following formula, we calculate the tone-mapped RGB values as a combination of the color information and a monochromatic intensity proportionally to the scotopic sensitivity:

$$\begin{bmatrix} R_L \\ G_L \\ B_L \end{bmatrix} = \begin{bmatrix} R \\ G \\ B \end{bmatrix} \cdot \frac{L(1 - \beta(Y))}{Y} + \begin{bmatrix} 1.05 \\ 0.97 \\ 1.27 \end{bmatrix} L\beta(Y), \tag{11.6}$$

where $[R_L, G_L, B_L]$ denotes the tone-mapped intensities, $[R, G, B]$ are the original HDR values, Y is the luminance, L is the tone-mapped luminance, and β is the scotopic sensitivity.

The constant coefficients in the monochromatic part account for the blue shift of the subjective hue of colors for night scenes.

11.3.5 Implementation

Global tone mapping operators require the computation of the global average of the luminance. When we have the luminance image we can calculate the global average with Gaussian filtering [47]. This can be a multi-step process, as we scale down the luminance image to one pixel in several passes. We can reduce the size of the luminance image in every step to reduce the computation.

The displayable luminance values are obtained at every pixel, and the scotopic vision formula is used to scale the RGB channels of the original HDR frame to displayable red, green, and blue values.

During the tone mapping process, in every pass we render a viewport-sized quadrilateral and let the fragment shader visit each texel. For example, the fragment shader computing the luminance for each pixel is

```
sampler2D HDRI; // source HDR image

float LuminancePS(in float2 Tex : TEXCOORD0) : COLOR {
   float3 col = tex2D(HDRI, Tex).rgb;
   return dot(col, float3(0.21, 0.71, 0.08));
}
```

In order to downscale the luminance image, the Gaussian filter is used, which is implemented according to the discussed importance sampling method (Section 11.1.3).

The final pass takes the average luminance value of the neighborhood and scales the color according to equations (11.2)–(11.6):

```
sampler2D AvgLumTex;   // Average luminance may be an 1x1 texture
float     beta0;       // Constant depending on the brightness of the screen

float4 FinalPS(float2 Tex : TEXCOORD0) : COLOR {
   float Ya = tex2D(AvgLumTex, Tex).r;            // Adaptation luminance
   float key = 1.03 - 2/(2 + log10(Ya + 1));      // equ. 11.4
   float Y = tex2D(LumTex, Tex).r;                // Luminance
   float Yr = key * Y / Ya;                       // equ. 11.2
   float L = Yr / (1 + Yr);                       // equ. 11.3
   float beta = beta0/(beta0 + Y);                // equ. 11.5
   float4 col = tex2D(HDRI, Tex) * L * (1 - beta)/Y +  // equ. 11.6
            float3(1.05,0.97,1.27) * L * beta;
   return pow(col, 1/2.2);                        // Gamma correction
}
```

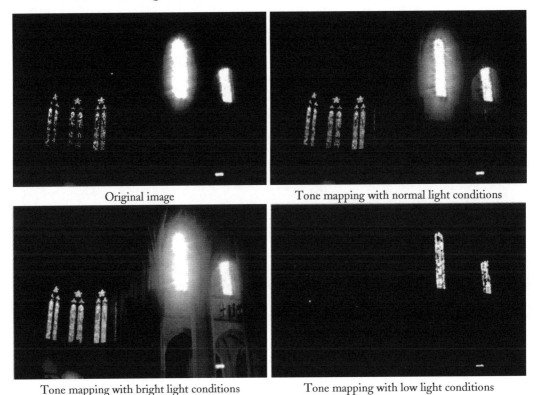

| Original image | Tone mapping with normal light conditions |

| Tone mapping with bright light conditions | Tone mapping with low light conditions |

FIGURE 11.4: Tone mapping results.

Note that the adaptation luminance is read from a texture. If global tone mapping is used, then this texture has just a single texel. In this case, it would be more efficient to compute the key factor only once instead of repeating the computation for every fragment.

11.4   DEPTH OF FIELD

Photorealistic rendering attempts to generate computer images with the quality approaching that of real life images. However, computer rendered images look synthetic or too perfect. Depth of field is one of those important visual components of real photography, which makes images look "real". In "real-world" photography, the physical properties of the camera cause some parts of the scene to be blurred, while maintaining sharpness in other areas being in focus. While blurriness sometimes can be thought of as an imperfection or undesirable artifact that distorts original images and hides some of the scene details, it can also be used as a tool to provide valuable visual clues and guide the viewer's attention to important parts of the scene. Depth of field may improve photorealism and add an artistic touch to rendered images.

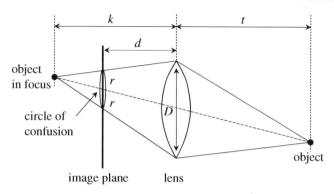

FIGURE 11.5: Image creation of real lens.

11.4.1 Camera Models and Depth of Field

Computer graphics generally implicitly uses the *pinhole camera model*, while real cameras have lenses of finite dimensions. The pinhole is in the eye position. In the pinhole camera, light rays scattered from objects pass through the infinitely small pinhole. Only a single ray emanating from each point in the scene is allowed to pass through. All rays going in other directions are ignored. Because only a single ray passes through the pinhole, only a single ray hits the imaging plane at any given point. This creates an image that is always in focus.

In the real world, however, all lenses have finite dimensions and let through rays coming from different directions. As a result, parts of the scene are sharp only if they are located at a specific focal distance. Let us denote the *focal length* of the lens by f, the diameter by D, and their ratio, called *aperture* number, by a:

$$a = \frac{f}{D}. \qquad (11.7)$$

The finite aperture causes the effect called *depth of field* which means that object points at a given distance appear in sharp focus on the image and other points beyond this distance or closer than that are confused, that is, they are mapped to finite extent patches instead of points.

It is known from geometric optics (see Figure 11.5) that if the focal length of a lens is f and an object point is at distance t from the lens, then the corresponding image point will be in sharp focus on an image plane at distance k behind the lens, where f, t, and k satisfy the following equation:

$$\frac{1}{f} = \frac{1}{k} + \frac{1}{t}. \qquad (11.8)$$

If the image plane is not at proper distance k, but at distance d as in Figure 11.5, then the object point is mapped onto a circle of radius r:

$$r = \frac{|k-d|}{k}\frac{D}{2} = \frac{|k-d|}{k}\frac{f}{2a}. \tag{11.9}$$

This circle is called the *circle of confusion* corresponding to the given object point. It expresses that the color of the object point affects the color of not only a single pixel but all pixels falling into the circle.

A given camera setting can be specified in the same way as in real life by the aperture number a and the *focal distance* P, which is the distance of those objects from the lens, which appear in sharp focus (not to be confused with the *focal length* of the lens). The focal distance and the distance of the image plane also satisfy the basic relation of the geometric optics:

$$\frac{1}{f} = \frac{1}{d} + \frac{1}{P}.$$

Subtracting equation (11.8) from this equation and using equation (11.9), we obtain the following formula for the radius of the circle of confusion:

$$r = \left| \frac{1}{t} - \frac{1}{P} \right| \frac{f}{2ad}.$$

According to this formula, the radius is proportional to the difference of the reciprocals of the object distance and of the focal distance. Since the projective transform and the homogeneous division translate camera space depth z to screen space depth Z as $Z = a + b/z$, where a and b depend on the front and back clipping space distances, the radius of the circle of confusion is just proportional to the difference of the object's depth coordinate and the focal distance, interpreting them in screen space.

11.4.2 Depth of Field With the Simulation of Circle of Confusion

The method presented in this section consists of two passes. In the first pass, the scene is rendered into textures of color, depth, and blurriness values. In the second pass, the final image is computed from the prepared textures with blurring.

The first pass can be accomplished by rendering the scene to multiple buffers at one time, using the *multiple render targets* (*MRT*) feature. One of the MRT restrictions on some hardware is the requirement for all render targets to have the same bit width, while allowing the use of different formats. Meeting this requirement we can pick the D3DFMT_A8R8G8B8 format for the scene color output and the D3DFMT_G16R16 format for depth and blurriness factors.

In the first pass, the fragment shader computes the blurriness factor and outputs it with the screen space depth and color. To abstract from the different display sizes and resolutions, the blurriness is defined in the [0, 1] range. A value of zero means the pixel is perfectly sharp, while a value of one corresponds to the maximal circle of confusion radius (maxCoc).

```
void DofVS(
      in  float4 Pos    : POSITION, // Input position in modeling space
      ...,
      out float4 hPos   : POSITION, // Output position in clipping space
      out float  hPos1  : TEXCOORD3 // Copy of hPos
) {
    hPos = mul(Pos, WorldViewProj);
    hPos1 = hPos;   // Copy clipping space position
    ...
}


void DofPS( in  float  hPos       : TEXCOORD3,// Clipping space position
            out float4 Color       : COLOR0,    // MRT: first target  -> ColorMap
            out float4 DepthBlur   : COLOR1     // MRT: second target -> DepthBlurMap
) {
    Color = ...;
    float depth = hPos.z / hPos.w;  // Screen space depth
    float blur  = saturate(abs(depth - focalDist) * focalRange);
    DepthBlur = float4(depth, blur, 0, 0);
}
```

During the second pass, the results of the first rendering pass are processed and the color image is blurred based on the blurriness factor computed in the first phase. Blurring is performed using a variable-sized filter representing the circle of confusion. To perform image filtering, a viewport-sized quad is drawn, which is textured with the results of the first phase. The filter kernel in the postprocessing step has 13 samples, containing a center sample and 12 outer samples. The postprocessing shader uses filter sample positions based on a 2D offset stored in filterTaps array. This array is initialized by the CPU with samples distributed with the Gaussian kernel according to the concept of *importance sampling* (Section 11.1.3). The array stores pairs $[\Phi^{-1}(r_1), \Phi^{-1}(r_2)]$ generated from uniformly distributed variables r_1, r_2 with the inverse of the cumulative distribution function of the Gaussian, $\Phi(x)$. Such an array can be initialized by the following CPU function:

```
void SetupFilterKernel() {
    float dx = 1.0 / HRES, {\rm d}y = 1.0 / VRES;
    D3DXVECTOR4 v[12];
    v[0]  = D3DXVECTOR4(-0.32621f * dx, -0.40581f * dy, 0.0f, 0.0f);
    v[1]  = D3DXVECTOR4(-0.84014f * dx, -0.07358f * dy, 0.0f, 0.0f);
```

```
    v[2]  = D3DXVECTOR4(-0.69591f * dx,  0.45714f * dy, 0.0f, 0.0f);
    v[3]  = D3DXVECTOR4(-0.20334f * dx,  0.62072f * dy, 0.0f, 0.0f);
    v[4]  = D3DXVECTOR4( 0.96234f * dx, -0.19498f * dy, 0.0f, 0.0f);
    v[5]  = D3DXVECTOR4( 0.47343f * dx, -0.48003f * dy, 0.0f, 0.0f);
    v[6]  = D3DXVECTOR4( 0.51946f * dx,  0.76702f * dy, 0.0f, 0.0f);
    v[7]  = D3DXVECTOR4( 0.18546f * dx, -0.89312f * dy, 0.0f, 0.0f);
    v[8]  = D3DXVECTOR4( 0.50743f * dx,  0.06443f * dy, 0.0f, 0.0f);
    v[9]  = D3DXVECTOR4( 0.89642f * dx,  0.41246f * dy, 0.0f, 0.0f);
    v[10] = D3DXVECTOR4(-0.32194f * dx, -0.93262f * dy, 0.0f, 0.0f);
    v[11] = D3DXVECTOR4(-0.79156f * dx, -0.59771f * dy, 0.0f, 0.0f);
    g_pPostEffect->SetVectorArray("filterTaps", v, 12);
}
```

One of the problems of postprocessing methods is the leaking of color from sharp objects onto the blurry background, which results in faint halos around sharp objects. This happens because the filter for the blurry background will sample color from the sharp object in the vicinity due to the large filter size. To solve this problem, we will discard samples that are in focus and are in front of the blurry center sample. This can introduce a minor popping effect when objects go in or out of focus. To combat sample popping, the outer sample blurriness factor is used as a sample weight to fade out its contribution gradually. The second pass renders a viewport-sized quad and executes the following fragment shader:

FIGURE 11.6: Depth of field with circle of confusion.

FIGURE 11.7: The Moria game with (left) and without (right) the depth of field effect.

```
const float maxCoC = 5;      // Maximum circle of confusion radius

float4 DepthBlurPS(in float2 Tex : TEXCOORD0) : COLOR {
    float4 colorSum = tex2D(ColorMapSampler, Tex);      // Color in the center sample
    float  depth = tex2D(DepthBlurMapSampler, Tex).r;   // Depth of the fragment
    float  blur  = tex2D(DepthBlurMapSampler, Tex).g;   // Circle of confusion radius
    float  sizeCoC = blur * maxCoC;  // CoC size based on blurriness

    float totalContrib = 1.0f;  // The center sample was already included
    for (int i = 0; i < NUM_DOF_TAPS; i++) { // Filter
        float2 tapCoord = Tex + filterTaps[i] * sizeCoC; // Tap coordinates
        // Fetch tap sample
        float4 tapColor = tex2D(ColorMapSampler, tapCoord);
        float  tapDepth = tex2D(DepthBlurMapSampler, tapCoord).r;
        float  tapBlur  = tex2D(DepthBlurMapSampler, tapCoord).g;
        // Compute tap contribution
        float tapContrib = (tapDepth > depth) ? 1 : tapBlur;
        colorSum += tapColor * tapContrib; // Accumulate color and contribution
        totalContrib += tapContrib;
    }
    return colorSum / totalContrib; // Normalize
}
```

CHAPTER 12

Integrating GI Effects in Games and Virtual Reality Systems

The previous chapters of this book discussed various global illumination methods and presented stand-alone programs and shaders that can implement them. In this last chapter, we examine how these approaches can be combined and integrated in *games* and *virtual reality* systems.

The integration of different global illumination effects poses both software engineering and graphics problems. From software engineering point of view, we should standardize the different solutions and develop a unified interface for them. This unified interface should meet the requirements of common game engines and scene graph management software that usually form the core of virtual reality systems. Since these real-time rendering systems are based on the local illumination paradigm, the dependency introduced by global illumination algorithms should be resolved somehow.

On the other hand, the integration of various global illumination approaches is a challenge from the graphics point of view as well. As more techniques are added, we wish to have better results, i.e. more accurate solutions of the rendering equation. The integration of a particular effect should extend the class of simulated light paths, but we must avoid including light paths of the same class more than once.

In the following section, we review the architecture of an advanced scene graph management software and discuss how its local illumination services can be used for the implementation of global illumination algorithms. Then another section is devoted to the graphics problems of the integration. Finally, case studies and integration examples are presented.

12.1 GAME ENGINES AND SCENE GRAPH MANAGERS

Game engines and virtual reality systems maintain a data structure that represents the current state of the *scene*, also called the *virtual world*. The scene is a collection of *scene objects*. This collection is heterogeneous since objects have different types, such as terrains, rigid bodies, static environment, characters animated with mesh morphing or with bones, particle systems, etc. This collection is also hierarchical since scene objects often have parent-child relationships.

For example, a weapon is the child of its owner since it usually follows him. Alternatively, a moon is the child of its planet since its own circular motion is relative to the planet's motion. The data structure representing the virtual world is a hierarchical, heterogeneous collection, implemented usually as a tree, which is usually called the *scene graph*. The scene graph stores both normal (e.g. an enemy character) and abstract objects (e.g. an abstract light source, or a viewport-sized quad). Abstract objects are not directly visible, but may affect other objects. The visibility of normal objects can be dynamically set.

The nodes of the scene graph have *bounding volumes* (e.g. spheres or boxes) that can contain the particular object and its children. Scene graph managers also maintain some *space partitioning data structure* (e.g. *bounding volume hierarchy, octree, kd-tree,* or *BSP-tree*) to quickly identify potentially visible objects, using, for example, *view frustum culling*[1]. Based on the space partitioning strategy, the scene graph manager can be asked to provide a *render queue* of the references of *potentially visible objects*. Note that the render queue may still store occluded objects, which are eliminated by the z-buffer hardware.

An object of the scene graph contains the definition of its geometry, usually as a set of triangle meshes, and transformations that scale, rotate, and translate the parts of this model. These motion parameters are set as a result of the physical simulation or by the artificial intelligence of the object. For the scope of this book, it is more important that in addition to geometric data, the object also has a *material* that describes the optical properties of its surface or volume. Of course, different objects may have the same material if we wish them to have a similar look.

If the visibility algorithm working with the geometry data and transformations concludes that a particular object is potentially visible in a pixel, then the information stored in the material is used to compute the color of this pixel. A material is primarily a collection of vertex, geometry, and fragment *shaders* that need to be activated in the GPU when an object having this material is sent down the graphics pipeline. A material may need several *passes*, where each pass has its own vertex, geometry, and fragment shader programs. If an object's material has more than one pass, then the scene graph manager will send this object multiple times down the pipeline having set the shaders of different passes one after the other. The results of the passes are merged according to the actual configuration of the *compositing* hardware (e.g. z-buffering, stencil testing, or alpha blending). If the program is prepared for various GPUs, the shader programs may have various versions appropriate for different GPU models.

---

[1]View frustum culling checks whether the cells of the space partitioning data structure or the bounding volumes of scene objects intersect the camera frustum. If not, the included objects cannot be visible, thus they can be skipped instead of sending them to the GPU for rendering.

A shader program may use *resources*. These resources can be 1D, 2D, or 3D textures in different formats, cube map textures, global parameters, constants, vertex buffer data, etc. The needed resource, e.g. a texture, can be uploaded by the CPU during the initialization of the scene or dynamically if some kind of *texture caching* approach is applied. Alternatively, the resource can also be generated by the GPU as an output of the pipeline, acting as a render target. In Shader Model 4.0 GPUs, the geometry shader may also output such resources.

The process of creating a resource is called the *rendering run*, which is the encapsulation of the generated output resource and additional parameters that control the shader programs. These additional parameters let the scene graph manager know which objects should be rendered, which materials, i.e. shaders should be activated while generating a certain resource, and where the camera should be placed. Concerning the materials, when an environment map is rendered, the material designed for the camera rendering can be used. However, when a shadow map is created, the color information of the objects is irrelevant. When generating a distance map, the fragment shader should also compute distance values in addition to the color information. In these cases, the original materials of the objects are replaced by materials associated with the pair of the original material and the resource (e.g. the shadow map or the distance map). Note that normally an object would have just a single material, but to handle the differences of shaders needed by rendering runs creating different resources, we need a collection of materials (Figure 12.1).

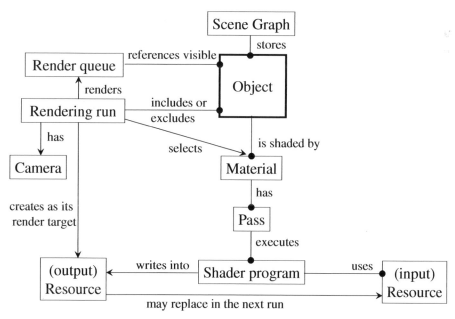

FIGURE 12.1: The model of shader management in scene graph software.

For the specification of which objects should be processed during a rendering run, the most general solution would be the listing of the references of objects for each rendering run. However, in practical cases a simpler solution also works. We distinguish *single object runs* and *multi object runs*. Single object runs render only one object that is referenced by the corresponding parameter of the rendering run. For example, a rendering run that executes tone mapping is a single object run since it would render just a single viewport-sized quad. Similarly, the creation of the distance map of a refractor, that stores normal vectors and is used for double refractions, is also a single object run since it requires the rendering of the refractor object into a cube map. Multi object runs, on the other hand, render all visible objects except the one that is referenced by the corresponding parameter. For example, the creation of an environment map requires the rendering of all objects but the reflector. If no object is excluded, then all visible objects are rendered, which is needed, for example, during the creation of a shadow map resource. The general rule applied by GPUs is that a resource generated in a rendering run can be used only in later rendering runs, so these resources are either readable or writable in a single run but not both.

A rendering run can be executed if its input resources are available, i.e. its input textures are either uploaded to the GPU memory or generated as the render target of a previous rendering run. This means that the scene graph manager is responsible for checking the validity of input resources before a particular rendering run is started. If an input resource is not available or obsolete, it should be created or updated before starting the rendering run that uses it.

The fact that resources needed by rendering may require other rendering runs introduces a dependency. The resolution of this dependency should be implemented very carefully since the creation of a resource may need other input resources, render targets, and shader programs. *Context switches* are expensive, thus their number should be minimized. A simple strategy would update all resources regularly and assume that they are always valid (because of regular updates this assumption is acceptable). On the other hand, we may detect the visibility of objects first using, for example, *potentially visible sets* [159], *view frustum culling* or *hierarchical occlusion culling* [157], and to test only those resources that are directly or even indirectly associated with at least one visible object. During a rendering run the potentially visible objects are put into the *render queue*. This queue should be sorted according to the required input resources and shaders to minimize the number of context switches.

If we consider the frame buffer also as a resource, then the global illumination rendering becomes equivalent to resolving *input resource* → *rendering run* → *output resource* dependences. Let us have a look at a few example dependences.

1. The *frame buffer resource* is generated from the *HDRI image resource* by the *postprocessing* rendering run, which may include *tone mapping*, *glow*, and *depth of field*, and requires the processing of a viewport-sized quad.

2. The *HDRI image resource* is the target of a rendering run that renders all potentially visible objects from the point of view of the camera. This requires that the input resources associated with these objects are valid. The input resources of shadow receivers are the shadow maps. A reflector or refractor needs distance maps. A caustic receiver needs the *light cube map* of the caustic generator, which in turn depends on the *photon hit location image* owned by the caustic generator and the light source.

3. The *shadow map* resource is created by a shadow map rendering run setting the shader programs to compute only the depth information, and the camera to look at the scene from the point of view of the light and to apply a distortion that minimizes shadow map aliasing.

4. A reflector, refractor or caustic generator can be rendered with its distance map. A distance map is created by the distance map rendering run, where all objects potentially visible from the center of the cube map are rendered except for the current object. Note that this rendering run may process other reflectors, that are rendered using their currently available distance map. This way *multiple reflections* between different objects are automatically presented, but *self reflections* are not.

The dependency graph may be complex and may also contain circles, so their absolutely correct resolution could be difficult or even impossible. For example, the cube map needed by the reflection computation on an object requires the color of another object. If the other object is also reflective, then its cube map generation may need the color of the first object (Section 6.3). However, if we update the resources frequently in (almost) every frame, then we can use the resource that has already been updated either in this frame, or in the last frame. The slight delays of indirect effects are not visible.

12.2 COMBINING DIFFERENT RENDERING ALGORITHMS

Different global illumination effects aim at tracing light paths of different types. The types of light paths are characterized by the notation proposed by Paul Heckbert [63]. The eye is denoted by E, the light source by L, a diffuse or glossy reflection on an optically rough surface by D, a specular, i.e. ideal reflection or refraction on an optically smooth surface by S, and light scattering in participating media by P. If scattering in participating media may only decrease the radiance but cannot change the direction of the light, i.e. only *forward scattering* is simulated, then the light attenuation is denoted by A. Note that radiance modifications due to diffuse or glossy reflection, specular reflection, and participating media scattering are represented by the BRDF, the Fresnel function, and by the pair of albedo and Phase function, respectively.

The history of events E, L, D, S, P, and A are expressed by sentences containing these events and symbols borrowed from the theory of formal languages. If an event is optional, then it is enclosed in []. If an event can happen one, two, etc., times, then it is included in {}. Events can also be ORed together with the symbol |.

The local illumination paradigm computes light paths of type LDE, i.e. only those light paths are considered, where the light emitted by a light source (L) is reflected once by a diffuse or glossy surface (D) toward the eye (E). Single scattering participating media rendering obtains $LP\{A\}E$ paths. On the other hand, the full solution of the global illumination problem should include light paths of arbitrary lengths and arbitrary reflection types, but still starting at the light source and terminating in the eye. Such paths are defined by sentence $L[\{D|S|P\}]E$. The goal of integrating different global illumination effects is to get closer to sentence $L[\{D|S|P\}]E$ without adding the contribution of a particular light path more than once.

The application of *shadow mapping* (Section 4.1) does not change the characterization of rendering algorithms. Shadow mapping improves the correctness of the geometric connection between light source L and light receiving surface D. *Image-based lighting* (Section 4.2) is a local illumination method, thus it is described by sentence LDE or LSE depending on whether or not the image has been convolved with the angular variation of the BRDF. *Environment mapping*, when the illumination hemisphere is computed as the image of the far environment, approximates $LDSE$ solutions (the rendered environment map represents D). Diffuse or glossy environment maps used to reflect the environment illumination on diffuse and glossy objects can help tracing $LDDE$ paths. Single specular reflections and refractions using distance maps of Chapter 6 also fall into the category of $LDSE$ paths, but correct the geometric errors made by environment mapping that assumes the environment surface to be infinitely far. Double refractions can result in $LDSSE$ paths, multiple reflections and refractions provide $LD\{S\}E$ ones. Of course, the number of followed specular reflections is always limited by the particular implementation. In caustic paths the light first meets specular surfaces then a diffuse surface, which can be expressed as $L\{S\}DE$. When caustics and specular reflections or refractions are combined (Section 6.8) $L\{S\}D\{S\}E$ paths are simulated.

Radiosity discussed in Section 7.1 may follow arbitrary number of diffuse reflections, and therefore is characterized by sentence $L[\{D\}]E$. The *diffuse and glossy final gathering* approach of Section 7.3 evaluates $LDDE$ paths. Comparing it to diffuse environment mapping, the difference is that the final gathering approach handles the distance between the object and the environment in a more accurate way.

Pre-computation aided approaches discussed in Chapter 8 handle mostly diffuse surfaces (it is possible to extend them for glossy surfaces as well, but the storage cost increases dramatically). However, they can handle not only surface reflections but also light scattering in participating media. *Pre-computed radiance transfer* (*PRT*), for example, stores light paths of type $\{D|P\}$ as real spherical harmonics series. Expressing the illumination in an orthogonal function base, $L\{D|P\}E$ paths can be dynamically obtained. *Light path maps* also represent

$D\{D\}$ paths, so they include only the indirect illumination (i.e. when at least two reflections happen). Direct illumination should be added to such solutions. The *illumination network* method of Section 9.6 pre-computes and stores paths of type $P\{P\}$. The participating media rendering method applying volume shadowing (Section 9.4) obtains $L[\{A\}]P\{A\}E$ paths. The *obscurances* method of Chapter 10 approximates indirect diffuse illumination, i.e. $L\{D\}DE$ paths, where L represents the ambient light source. Postprocessing techniques do not change the type of light paths. Instead they simulate operator E in a more realistic way than the ideal pinhole camera model would do.

The combination of different techniques should avoid the inclusion of the same light paths more than once. Those techniques that generate disjoint sets of light paths can be safely combined and are said to be *compatible*. However, techniques generating non-disjoint light path sets require special considerations. For example, a standard local illumination algorithm (LDE) cannot be directly combined on the same object with methods that also compute LDE paths, for example, with radiosity ($L[\{D\}]E$) or PRT ($L\{D|P\}E$). However, the combination is feasible if a subset of light sources are taken into account by direct illumination while other light sources are handled by radiosity or PRT. This is a standard approach in games, where static area lights are processed by radiosity simulating multiple reflections, while dynamic point or directional lights are processed on the fly but only their direct illumination is computed. Concerning PRT we can decide for each object–light source pair whether it is processed by local illumination or by PRT. Note that PRT involves also self-shadowing computation.

The combination of direct illumination and techniques computing only the indirect illumination, such as the diffuse and glossy final gathering or the light path maps, poses no problem, the radiance values of the two techniques should be simply added.

It makes no sense to combine environment mapping and the distance map-based reflection algorithm since both of them aim at the same phenomenon but with different accuracy. We should select from these alternatives according to the expected accuracy or parallax effects and the available hardware. Caustics are unique in the sense that the light of the light source should arrive at a specular surface first, i.e. its sentence starts with LS. Since none of the other techniques is able to simulate such paths, caustic patterns can be safely added to the image computed by other methods.

The selection of the participating media rendering algorithm to be integrated is based on the expected accuracy and rendering speed. Illumination networks are more accurate, but require pre-computation, which limits their application to static phenomena. If the media are dynamic, the cheaper volume shadowing (forward scattering) algorithm is the feasible alternative.

Although integrating compatible techniques improves the accuracy of the global illumination rendering, we cannot expect our program to deliver a fully accurate global illumination solution. The presented methods as well as their integrated versions make compromises in

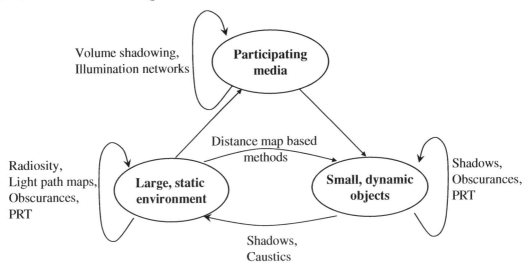

FIGURE 12.2: A simplified global illumination model for games.

the accuracy to guarantee real-time frame rates. These compromises lead us to the simplified global illumination model of Figure 12.2, which also shows the roles of particular techniques. According to this model the scene is decomposed to a static environment (e.g. city, terrain, building, etc.) and to dynamic objects (e.g. enemy character, vehicle, weapon, etc.), and the static environment is assumed to be significantly larger. Due to these size differences, the indirect illumination of the static environment on dynamic objects is important, but dynamic objects do not significantly modify the indirect illumination of the static environment. There is one exception, the caustic phenomenon, which is very intensive and cannot be omitted. On the other hand, dynamic objects do modify the direct illumination of the static environment, which is solved by shadow mapping.

The self illumination of the static environment can be pre-computed and stored as radiosity textures, conventional light maps, spherical harmonics coefficients used by PRT, and light path maps. Radiosity textures and light maps use fixed light source positions, while PRT and light path maps allow the animation of light sources. During pre-computation dynamic objects are ignored. Since they are small, we can assume that they would not alter the indirect illumination too much.

The indirect illumination of the static environment on dynamic objects can be efficiently computed by classical environment mapping or by distance map-based techniques.

12.3   CASE STUDIES

Current games usually pre-compute radiosity textures for the static environment and handle dynamic objects with some shadow algorithm. Reflective and refractive objects are processed

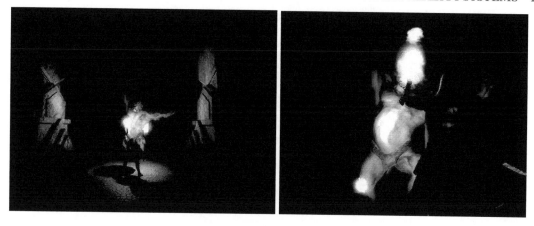

FIGURE 12.3: A knight and a troll are fighting in Moria illuminated by the light path map algorithm.

with environment mapping. Participating media is visualized by assuming only forward scattering. In this section we review three demo games that stepped beyond this state of the art. The demo games have been implemented using a modified version of the *Ogre3D* engine (http://www.ogre3d.org).

12.3.1 Moria

The scene of Moria consists of large, dark, and bright colonnades, where trolls and knights fight (Figure 12.3). The halls are sometimes foggy or full of smoke. One hall has a bright light source, but the other room is illuminated only indirectly through the door, which makes the lighting very uneven and similar to real scenes (Figure 12.4). The illumination is distributed in the scene by *light path maps*. To handle large variations of the light intensity the scene is rendered first to a HDRI image, from which the frame buffer image is obtained with *tone mapping*. We implemented the tone mapping operator of Section 11.3, that simulates temporal adaptation and also the glowing of bright objects. Additionally the depth of field effect was also added, assuming that the user looks at the center of the screen. Objects in Moria are mostly diffuse or glossy with the exception of the knight's shield and sword. To render these specular, metallic objects, we used the distance map-based ray-tracing algorithm of Chapter 6, and the Fresnel approximation of equation (1.6) having looked up the index of refraction and the extinction coefficient of steal from physics textbooks at the wavelengths of red, green, and blue. The knight's armor is glossy, which also reflects its environment. This effect is computed with the algorithm of Section 7.3. The diffuse version of the same algorithm is used to get the indirect illumination of the troll. The shadows are computed by the *variance shadow map* algorithm.

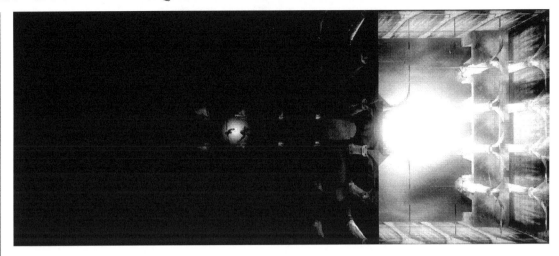

FIGURE 12.4: A bird's eye view of the Moria scene. Note that the hall on the right is illuminated by a very bright light source. The light enters the hall on the left after multiple reflections computed by the light path map method. The troll and the knight with his burning torch can be observed in the middle of the dark hall.

The dynamic smoke is rendered using the *spherical billboard* method and the volume illumination algorithm of Section 9.4. The results are shown in Figure 12.6. Note that spherical billboards successfully eliminated billboard clipping artifacts. In addition to the sword, the knight has another weapon, the fireball (Figure 12.7). The dynamic fire is created by the fire algorithm of Section 9.5, and also acts as a light source.

FIGURE 12.5: Tone mapping with glow (left) and depth of field (right) in Moria.

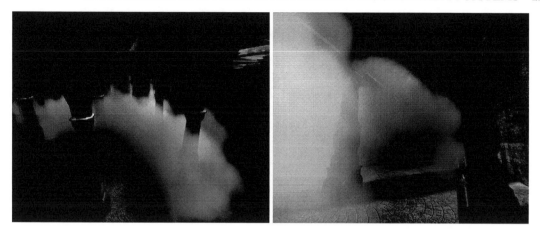

FIGURE 12.6: Shaded smoke in Moria with spherical billboards and simplified multiple scattering.

12.3.2 RT Car

The RT car game demonstrates the distance map-based *specular* effects, *diffuse and glossy final gathering*, and also the *explosion* effect. RT stands for ray-trace and also for real-time, and indicates that distance map-based approximate ray tracing can really produce specular effects in real-time environments.

The car has a glossy + specular body and a diffuse roof (Figures 12.8 and 12.9) that reflect their local environment. The wheels of the car are specular and generate caustics. In the first arena of the game, the car is expected to push giant beer bottles that are made of colored glass. The beer bottles are reflective, refractive, and also caustic generators. The scene is illuminated by the sky (we use *image-based lighting*) and by the Sun, which is a strong directional light.

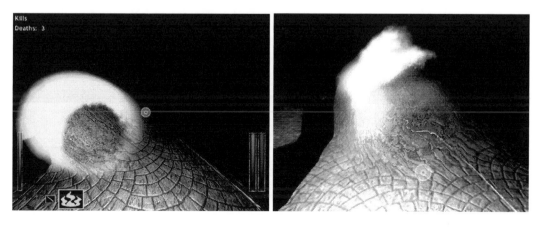

FIGURE 12.7: Fireball causing heat shimmering in Moria.

FIGURE 12.8: Snapshots from the RT car game. Note that the car body specularly reflects the environment. The wheels are not only reflectors but are also caustic generators. The caustics are clearly observable on the ground. The giant beer bottles reflect and refract the light and also generate caustics. Since the car and the bottle have their own distance maps, the inter-reflections between them are also properly computed.

   We assigned a distance map for the car body, one for the two front wheels, one for the two rear wheels, and one for each beer bottle. When the color channels of a distance map are rendered, the distance map-based reflection calculation of other objects is executed. This way we can render *multiple specular effects*, for example, the refraction on the beer bottle of the car, which in turn reflects the arena. However, this approach is not able to render self reflections

FIGURE 12.9: The car not only specularly but also diffusely reflects the indirect illumination, which is particularly significant when the car is in the corridor.

FIGURE 12.10:   Explosions with heat shimmering in RT car demo game.

of the objects (self reflections would need the algorithm of Section 6.6, which we have not implemented in this game). The two front wheels cannot reflect each other, thus it is not worth assigning a separate cube map to them, but can share the same light cube map. This idea of *resource sharing* can be generalized and we can replace two or more light cube maps by a single one anytime their owners get close.

The second arena of this game stores gas tanks. Hitting these tanks by the car, *explosions* are created that also produce *heat shimmering* (Figure 12.10). We used the explosion effect of Section 9.5. Since the gas tanks remain there during explosions, the spherical billboards are

FIGURE 12.11:   Indirect diffuse illumination in the Space Station game. The self illumination of the space station is computed by the light path maps method. The indirect illumination of the space station onto the scientist character is obtained by diffuse or glossy final rendering. The eyeglasses of the character specularly reflects the environment. Shadows are computed with the variance shadow map algorithm.

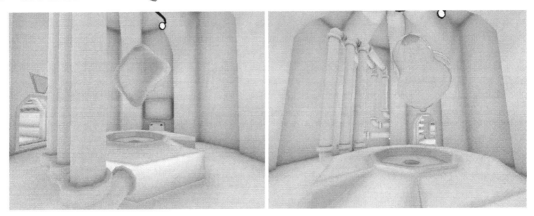

FIGURE 12.12:   The space station rendered with scalar obscurances only.

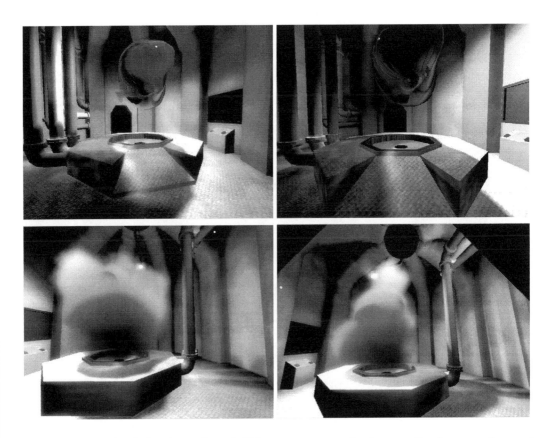

FIGURE 12.13: A deforming glass bubble generates reflections, refractions, and caustics (upper row). The bubble is surrounded by smoke (lower row).

badly needed to eliminate clipping artifacts of classical billboards. Simultaneously, the car is bounced back as simulated by the rigid body physics solver.

The arenas are connected by corridors that have colored walls and roofs, which significantly modify the Sun's and the sky's illumination. To cope with the modified indirect illumination of the local environment, we compute distance map-based specular and diffuse or glossy reflections (Figures 12.8 and 12.9).

12.3.3 Space Station

The space station is a labyrinth of corridors and rooms. Only the rooms have dynamic, moving light sources, the corridors are only illuminated by the indirect lighting of the rooms and by the torch in the hand of the avatar if it is turned on. Note that we do not use ambient lighting in the space station (similarly to Moria). The rooms and corridors are lit by the *light path map* global illumination algorithm that can also handle dynamic lights and produces appealing color bleeding effects (Figure 12.11). The dynamic characters are mainly diffuse (e.g. the scientist) and are illuminated by their environments using the distance map-based specular and diffuse or glossy reflections. The scientist has specular eyeglasses.

The texture memory needed by light path maps can be reduced by the *clustering* algorithm discussed in Section 8.4.1 or by decreasing the resolution of these maps, but these would filter higher frequency details of the indirect illumination. A simple trick to re-introduce these higher frequency details is the application of scalar obscurances (Figure 12.13) and modulate the indirect illumination of light path maps by the fetched scalar obscurance.

The space station also has a dynamic specular object, a deforming glass bubble, which reflects, refracts light, and generates caustics of the moving light sources (Figure 12.13). The user can also turn on and off a smoke generator, which emits smoke surrounding the glass bubble. The smoke is illuminated by the algorithm of Section 9.4.

Bibliography

[1] J. Amanatides and A. Woo, "A fast voxel traversal algorithm for ray tracing," in *Proc. Eurographics '87*, pp. 3–10, 1987.

[2] T. Annen, J. Kautz, F. Durand, and H.-P. Seidel, "Spherical harmonic gradients for mid-range illumination," in *Eurographics Symp. on Rendering*, 2004.

[3] Gy. Antal and L. Szirmay-Kalos, "Fast evaluation of subdivision surfaces on Direct3D 10 graphics hardware," in Wolfgang Engel, editor, *ShaderX6: Advanced Rendering Techniques*. Boston, MA: Charles River Media, 2007.

[4] A. A. Apodaca and L. Gritz, *Advanced RenderMan: Creating CGI for Motion Picture*. San Francisco: Morgan Kaufman Publishers, Inc., 1999.

[5] J. Arvo, "Backward ray tracing," in *SIGGRAPH '86 Developments in Ray Tracing*, 1986.

[6] J. Arvo, "Linear-time voxel walking for octrees," *Ray Tracing News*, 1(2), 1991, available under anonymous ftp from weedeater.math.yale.edu.

[7] J. Arvo and D. Kirk, "Fast ray tracing by ray classification," in *Computer Graphics (SIGGRAPH '87 Proc.)*, pp. 55–64, 1987.

[8] J. Arvo and D. Kirk, "A survey of ray tracing acceleration techniques," in Andrew S. Glassner, editor, *An Introduction to Ray Tracing*, pp. 201–262. London: Academic Press, 1989.

[9] L. Aupperle and P. Hanrahan, "A hierarchical illumination algorithms for surfaces with glossy reflection," *Computer Graphics (SIGGRAPH '93 Proc.)*, pp. 155–162, 1993.

[10] A. Barsi, L. Szirmay-Kalos, and L. Szécsi, "Image-based illumination on the GPU," *Mach. Graph. Vis.*, 14(2):159–170, 2005.

[11] A. Barsi, L. Szirmay-Kalos, and G. Szijártó, "Stochastic glossy global illumination on the GPU," in *Proc. Spring Conf. on Computer Graphics (SCCG 2005)*, Slovakia: Comenius University Press.

[12] R. Bastos, M. Goslin, and H. Zhang, "Efficient radiosity rendering using textures and bicubic reconstruction," in *ACM-SIGGRAPH Symp. on Interactive 3D Graphics*, 1997.

[13] R. Bastos, K. Hoff, W. Wynn, and A. Lastra, "Increased photorealism for interactive architectural walkthroughs," in *SI3D 99: Proc. 1999 Symp. on Interactive 3D Graphics*, pp. 183–190, 1999.

[14] P. Bekaert, "Hierarchical and stochastic algorithms for radiosity," Ph.D. thesis, University of Leuven, 1999.

[15] K. Bjorke, "Image-based lighting," in R. Fernando, editor, *GPU Gems*, pp. 307–322. MA: Addison-Wesley, 2004.

[16] J. F. Blinn and M. E. Newell, "Texture and reflection in computer generated images," *Commun. ACM*, 19(10):542–547, 1976.

[17] S. Brabec, T. Annen, and H.-P. Seidel, "Practical shadow mapping," *J. Graph. Tools*, 7(4):9–18, 2002.

[18] M. Bunnel, "Dynamic ambient occlusion and indirect lighting," in M. Parr, editor, *GPU Gems 2*, pp. 223–233. Reading, MA: Addison-Wesley, 2005.

[19] N. Carr, J. Hall, and J. Hart, "The ray engine," in *Proc. Graphics Hardware*, 2002.

[20] N. Carr, J. Hall, and J. Hart, "GPU algorithms for radiosity and subsurface scattering," in *Proc. Workshop on Graphics Hardware*, pp. 51–59, 2003.

[21] N. Carr, J. Hoberock, K. Crane, and J. Hart, "Fast GPU ray tracing of dynamic meshes using geometry images," in *Proc. Graphics Interface*, pp. 203–209, 2006.

[22] M. Chen and J. Arvo, "Perturbation methods for interactive specular reflections," *IEEE Trans. Vis. Comput. Graph.*, 6(3):253–264, 2000.

[23] M. Christen, "Ray tracing on GPU," Master's thesis, University of Applied Sciences, Basel (FHBB), 2005.

[24] M. Cohen, S. Chen, J. Wallace, and D. Greenberg, "A progressive refinement approach to fast radiosity image generation," in *Computer Graphics (SIGGRAPH '88 Proc.)*, pp. 75–84, 1988.

[25] M. Cohen and D. Greenberg, "The hemi-cube, a radiosity solution for complex environments," in *Computer Graphics (SIGGRAPH '85 Proc.)*, pp. 31–40, 1985.

[26] G. Coombe, M. J. Harris, and A. Lastra, "Radiosity on graphics hardware," in *Proc. Graphics Interface*, 2004.

[27] W. Cornette and J. Shanks, "Physically reasonable analytic expression for single-scattering phase function," *Appl. Opt.*, 31(16):31–52, 1992.

[28] F. Crow, "Shadow algorithm for computer graphics," in *Computer Graphics (SIGGRAPH '77 Proc.)*, pp. 242–248, 1977.

[29] C. Dachsbacher and M. Stamminger, "Reflective shadow maps," in *SI3D '05: Proc. 2005 Symp. on Interactive 3D Graphics and Games*, pp. 203–231, 2005.

[30] P. Debevec, "Rendering synthetic objects into real scenes: Bridging traditional and image-based graphics with global illumination and high dynamic range photography," in *SIGGRAPH '98*, pp. 189–198, 1998.

[31] P. Debevec, G. Borshukov, and Y. Yu, "Efficient view-dependent image-based rendering with projective texture-mapping," in *9th Eurographics Rendering Workshop*, 1998.

[32] P. Debevec and J. Malik, "Recovering high dynamic range radiance maps from photographs," in *SIGGRAPH '97*, 1997.

[33] P. Diefenbach and N. Badler, "Multi-pass pipeline rendering: realism for dynamic environments," in *SI3D 97: Proc. 1997 Symposium on Interactive 3D Graphics*, pp. 59–68, 1997.

[34] W. Donnelly and A. Lauritzen, "Variance shadow maps," in *SI3D '06: Proc. 2006 Symposium on Interactive 3D Graphics and Games*, pp. 161–165.

[35] P. Dutre, P. Bekaert, and K. Bala, *Advanced Global Illumination*. Wellesley, MA: A K Peters, 2003.

[36] M. Ernst, C. Vogelgsang, and G. Greiner, "Stack implementation on programmable graphics hardware," in *Proc. Vision, Modeling, and Visualization*, pp. 255–262, 2004.

[37] P. Estalella, I. Martin, G. Drettakis, and D. Tost, "A GPU-driven algorithm for accurate interactive specular reflections on curved objects," in *Proc. 2006 Eurographics Symp. on Rendering*.

[38] P. Estalella, I. Martin, G. Drettakis, D. Tost, O. Devilliers, and F. Cazals, "Accurate interactive specular reflections on curved objects," in *Proc. VMV*, 2005.

[39] C. Everitt, "Interactive order-independent transparency," Technical report, NVIDIA Corporation, 2001.

[40] J. Evers-Senne and R. Koch, "Image based interactive rendering with view dependent geometry," *Eurographics*, 22(3):573–582, 2003.

[41] T. Foley and J. Sugerman, "Kd-tree acceleration structures for a GPU raytracer," in *Proc. Graphics Hardware*, 2005.

[42] A. Fujimoto, T. Takayuki, and I. Kansei, "Arts: accelerated ray-tracing system," *IEEE Comput. Graph. Appl.*, 6(4):16–26, 1986.

[43] M. Giegl and M. Wimmer, "Fitted virtual shadow maps," in *Proc. Graphics Interface*, 2007.

[44] A. Glassner, "Space subdivision for fast ray tracing," *IEEE Comput. Graph. Appl.*, 4(10):15–22, 1984.

[45] A. Glassner, *An Introduction to Ray Tracing*. London: Academic Press, 1989.

[46] A. Glassner, *Principles of Digital Image Synthesis*. San Francisco: Morgan Kaufmann Publishers, Inc., 1995.

[47] N. Goodnight, R. Wang, C. Woolley, and G. Humphreys, "Interactive time-dependent tone mapping using programmable graphics hardware," in *Rendering Techniques 2003: 14th Eurographics Symp. on Rendering*, pp. 26–37.

[48] S. Gortler, R. Grzeszczuk, R. Szeliski, and M. Cohen, "The lumigraph," in *SIGGRAPH 96*, pp. 43–56, 1996.

[49] C. Green, "Efficient self-shadowed radiosity normal mapping," in *ACM SIGGRAPH 2007. Course 28: Advanced Real-time Rendering in 3D Graphics and Games.*

[50] R. Green, "Spherical harmonic lighting: the gritty details," Technical report, 2003. http://www.research.scea.com/gdc2003/ spherical-harmonic-lighting.pdf.

[51] N. Greene, "Environment mapping and other applications of world projections," *IEEE Comput. Graph. Appl.*, 6(11):21–29, 1984.

[52] T. Hachisuka, "High-quality global illumination rendering using rasterization," in M. Parr, editor, *GPU Gems 2*. Reading, MA: Addison-Wesley, 2005, pp. 615–634.

[53] M. Hadwiger, C. Sigg, H. Scharsach, K. Bühler, and M. Gross, "Real-time ray-casting and advanced shading of discrete isosurfaces," in *Eurographics Conference, 2005*, pp. 303–312.

[54] P. Hanrahan, D. Salzman, and L. Aupperle, "Rapid hierarchical radiosity algorithm," *Computer Graphics (SIGGRAPH '91 Proc.)*, 1991.

[55] M. Harris, "Real-time cloud rendering for games," in *Game Developers Conference*, 2002.

[56] M. Harris, W. Baxter, T. Scheuermann, and A. Lastra, "Simulation of cloud dynamics on graphics hardware," in *Eurographics Graphics Hardware 2003*.

[57] M. Harris and A. Lastra, "Real-time cloud rendering," *Comput. Graph. Forum*, 20(3), 2001.

[58] J.-M. Hasenfratz, M. Lapierre, N. Holzschuch, and F. X. Sillion, "A survey of realtime soft shadow algorithms," in *Eurographics Conference. State of the Art Reports*, 2003.

[59] V. Havran, *Heuristic Ray Shooting Algorithms*. Czech Technical University, Ph.D. dissertation, 2001.

[60] V. Havran, R. Herzog, and H.-P. Seidel, "On the fast construction of spatial hierarchies for ray tracing," in *Proc. IEEE Symp. on Interactive Ray Tracing*, pp. 71–80, 2006.

[61] V. Havran and W. Purgathofer, "On comparing ray shooting algorithms," *Comput. Graph.*, 27(4):593–604, 2003.

[62] L. Hayden, "Production-ready global illumination," SIGGRAPH Course notes 16, 2002. http://www.renderman.org/RMR/Books/ sig02.course16.pdf.gz.

[63] P. Heckbert, "Simulating global illumination using adaptive meshing," Ph.D. thesis. University of California, Berkeley, 1991.

[64] W. Heidrich, H. Lensch, M. Cohen, and H.-P. Seidel, "Light field techniques for reflections and refractions," in *Eurographics Rendering Workshop*, 1999.

[65] G. Henyey and J. Greenstein, "Diffuse radiation in the galaxy," *Astrophys. J.*, 88:70–73, 1940.

[66] J. Hirche, A. Ehlert, S. Guthe, and M. Doggett, "Hardware accelerated per-pixel displacement mapping," in *Proc. Graphics Interface*, pp. 153–158, 2004.

[67] A. Iones, A. Krupkin, M. Sbert, and S. Zhukov, "Fast realistic lighting for video games," *IEEE Comput. Graph. Appl.*, 23(3):54–64, 2003.

[68] H. W. Jensen, "Global illumination using photon maps," in *Rendering Techniques '96*, pp. 21–30, 1996.

[69] H. W. Jensen and P. H. Christensen, "Photon maps in bidirectional Monte Carlo ray tracing of complex objects," *Comput. Graph.*, 19(2):215–224, 1995.

[70] H. W. Jensen and P. H. Christensen, "Efficient simulation of light transport in scenes with participating media using photon maps," *SIGGRAPH '98 Proc.*, pp. 311–320, 1998.

[71] H. W. Jensen, S. Marschner, M. Levoy, and P. Hanrahan, "A practical model for subsurface light transport," *Computer Graphics (SIGGRAPH 2001 Proc.)*, 2001.

[72] J. T. Kajiya, "The rendering equation," in *Computer Graphics (SIGGRAPH '86 Proc.)*, pp. 143–150, 1986.

[73] T. Kanungo, D. Mount, N. Netanyahu, C. Piatko, R. Silverman, and A. Wu, "An efficient k-means clustering algorithm: analysis and implementation," *IEEE Trans. Pattern Anal. Mach. Int.*, 24(7):881–892, 2002.

[74] F. Karlsson and C. J. Ljungstedt, "Ray tracing fully implemented on programmable graphics hardware," Master's thesis, Chalmers University of Technology, 2004.

[75] J. Kautz and M. McCool, "Approximation of glossy reflection with prefiltered environment maps," in *Proc. Graphics Interface*, 2000.

[76] J. Kautz, P. Sloan, and J. Snyder, "Fast, arbitrary BRDF shading for low-frequency lighting using spherical harmonics," in *12th EG Workshop on Rendering*, pp. 301–308, 2002.

[77] J. Kautz, P. Vázquez, W. Heidrich, and H.-P. Seidel, "A unified approach to prefiltered environment maps," in *11th Eurographics Workshop on Rendering*, pp. 185–196, 2000.

[78] A. Keller, "Instant radiosity," in *SIGGRAPH '97 Proc.*, pp. 49–55, 1997.

[79] G. King, "Real-time computation of dynamic irradiance environment maps," in M. Parr, editor, *GPU Gems 2*. Reading, MA: Addison-Wesley, 2005, pp. 167–176.

[80] T. Kollig and A. Keller, "Efficient illumination by high dynamic range images," in *Eurographics Symp. on Rendering*, pp. 45–51, 2003.

[81] J. Kontkanen and T. Aila, "Ambient occlusion for animated characters," in *Proc. 2006 Eurographics Symp. on Rendering*.

[82] G. Krawczyk, K. Myszkowski, and H.-P. Seidel, "Perceptual effects in real-time tone mapping," in *SCCG '05: Proc. 21st Spring Conf. on Computer Graphics*. New York, NY: ACM Press, 2005, pp. 195–202.

[83] A. W. Kristensen, T. Akenine-Moller, and H. W. Jensen, "Precomputed local radiance transfer for real-time lighting design," in *SIGGRAPH*, 2005.

[84] J. Krüger, K. Bürger, and R. Westermann, "Interactive screen-space accurate photon tracing on GPUs," in *Eurographics Symp. on Rendering*, 2006.

[85] I. Lazányi and L. Szirmay-Kalos, "Fresnel term approximations for metals," in *WSCG 2005, Short Papers*, pp. 77–80.

[86] J. Lehtinen and J. Kautz, "Matrix radiance transfer," in *SI3D '03: Proc. 2003 Symp. on Interactive 3D Graphics*, pp. 59–64.

[87] H. Lensch, M. Goesele, Ph. Bekaert, J. Kautz, M. Magnor, J. Lang, and H.-P. Seidel, "Interactive rendering of translucent objects," *Comput. Graph. Forum*, 22(2):195–195, 2003.

[88] M. Levoy and P. Hanrahan, "Light field rendering," in *SIGGRAPH 96*, pp. 31–42, 1996.

[89] B. Li, L.-Y. Wei, and Y.-Q. Xu, "Multi-layer depth peeling via fragment sort," Technical report MSR-TR-2006-81, Microsoft Research, 2006.

[90] D. Lischinski and A. Rappoport, "Image-based rendering for non-diffuse synthetic scenes," in *Eurographics Rendering Workshop*, pp. 301–314, 1998.

[91] S. Lloyd, "Least square quantization in pcm," *IEEE Trans. Inform. Theory*, 28:129–137, 1982.

[92] A. Mendez, M. Sbert, J. Cata, N. Sunyer, and S. Funtane, "Real-time obscurances with color bleeding," in Wolfgang Engel, editor, *ShaderX4: Advanced Rendering Techniques*. Boston, MA: Charles River Media, 2005.

[93] J. Meseth, M. Guthe, and R. Klein, "Interactive fragment tracing," *J. Vis. Comput.*, 21(8–10):591–600, 2005.

[94] L. Neumann, "Monte Carlo radiosity," *Computing*, 55:23–42, 1995.

[95] R. Ng, R. Ramamoorthi, and P. Hanrahan, "All-frequency shadows using non-linear wavelet lighting approximation," *ACM Trans. Graph.*, 22(3):376–381, 2003.

[96] H. Nguyen, "Fire in the vulcan demo," in R. Fernando, editor, *GPU Gems*, pp. 359–376. Reading, MA: Addison-Wesley, 2004.

[97] H. Niederreiter, *Random Number Generation and Quasi-Monte Carlo Methods*. Pennsilvania: SIAM, 1992.

[98] K. Nielsen and N. Christensen, "Fast texture based form factor calculations for radiosity using graphics hardware," *J. Graphics Tools*, 6(2):1–12, 2002.

[99] K. Nielsen and N. Christensen, "Real-time recursive specular reflections on planar and curved surfaces using graphics hardware," *J. WSCG*, 10(3):91–98, 2002.

[100] E. Ofek and A. Rappoport, "Interactive reflections on curved objects," in *Proc. SIGGRAPH*, pp. 333–342, 1998.

[101] M. Ohta and M. Maekawa, "Ray coherence theorem and constant time ray tracing algorithm," in T. L. Kunii, editor, *Computer Graphics 1987. Proc. CG International '87*, pp. 303–314, 1987.

[102] V. Ostromoukhov, C. Donohue, and P-M. Jodoin, "Fast hierarchical importance sampling with blue noise properties," in *Proc. SIGGRAPH*, 2004.

[103] J. Owens, D. Luebke, N. Govindaraju, M. Harris, J. Krüger, A. Lefohn, and T. Purcell, "A Survey of General-Purpose Computation on Graphics Hardware," in *EG2005-STAR*, pp. 21–51, 2005.

[104] M. Parr and S. Green, "Ambient occlusion," in *GPU Gems*, pp. 279–292. Reading, MA: Addison-Wesley, 2004.

[105] G. Patow, "Accurate reflections through a z-buffered environment map," in *Proc. Sociedad Chilena de Ciencias de la Computacion*, 1995.

[106] V. Popescu, C. Mei, J. Dauble, and E. Sacks, "Reflected-scene impostors for realistic reflections at interactive rates," *Comput. Graph. Forum (Eurographics 2006)*, 25(3).

[107] A. Preetham, P. Shirley, and B. Smits, "A practical analytic model for daylight," in *SIGGRAPH '99 Proccedings*, pp. 91–100, 1999.

[108] K. Pulli, M. Cohen, T. Dunchamp, H. Hoppe, L. Shapiro, and W. Stuetzle, "View-based rendering: Visualizing real objects from scanned range and color data," in *8th Eurographics Rendering Workshop*, pp. 23–34, 1997.

[109] T. Purcell, I. Buck, W. Mark, and P. Hanrahan, "Ray tracing on programmable graphics hardware. *ACM Trans. Graph.*, 21(3):703–712, 2002.

[110] T. Purcell, T. Donner, M. Cammarano, H. W Jensen, and P. Hanrahan, "Photon mapping on programmable graphics hardware," in *Proc. ACM SIGGRAPH/EUROGRAPHICS Conf. on Graphics Hardware*, pp. 41–50, 2003.

[111] R. Ramamoorthi and P. Hanrahan, "An efficient representation for irrandiance environment maps," *SIGGRAPH 2001*, pp. 497–500.

[112] W. T. Reeves, D. H. Salesin, and R. L. Cook, "Rendering antialiased shadows with depth maps," pp. 283–291, 1987.

[113] E. Reinhard, "Parameter estimation for photographic tone reproduction," *J. Graphics Tools: JGT*, 7(1):45–52, 2002.

[114] E. Reinhard, L. U. Tijssen, and W. Jansen, "Environment mapping for efficient sampling of the diffuse interreflection," in *Photorealistic Rendering Techniques*. Berlin: Springer, 1994, pp. 410–422.

[115] E. Reinhard, G. Ward, S. Pattanaik, and P. Debevec, *High Dynamic Range Imaging*. San Francisco: Morgan Kaufmann, 2006.

[116] A. Reshetov, A. Soupikov, and J. Hurley, "Multi-level ray tracing algorithm," *ACM Trans. Graph.*, 24(3):1176–1185, 2005.

[117] K. Riley, D. Ebert, M. Kraus, J. Tessendorf, and C. Hansen, "Efficient rendering of atmospheric phenomena," in *Eurographics Symp. on Rendering*, pp. 374–386, 2004.

[118] D. Roger, N. Holzschuch, and F. Sillion, "Accurate specular reflections in real-time," *Comput. Graph. Forum (Eurographics 2006)*, 25(3).

[119] P. Sander, "DirectX9 High Level Shading Language," in *Siggraph 2005 Tutorial*, 2005. http://ati.amd.com/developer/SIGGRAPH05/ ShadingCourse_HLSL.pdf.

[120] M. Sbert, "The use of global directions to compute radiosity," Ph.D. thesis, Catalan Technical University, Barcelona, 1997.

[121] M. Sbert, L. Szécsi, and L. Szirmay-Kalos, "Real-time light animation," *Comput. Graph. Forum (Eurographics 2004)*, 23(3):291–300.

[122] H. Schirmacher, L. Ming, and H.-P. Seidel, "On-the-fly processing of generalized lumigraphs," *Eurographics*, 20(3):165–173, 2001.

[123] Ch. Schlick, "A customizable reflectance model for everyday rendering," in *Fourth Eurographics Workshop on Rendering*, pp. 73–83, 1993.

[124] M. Shah and S. Pattanaik, "Caustics mapping: An image-space technique for real-time caustics," *IEEE Trans. Vis. Comput. Graph. (TVCG)*, March 2006.

[125] P. Shirley, "Time complexity of Monte-Carlo radiosity," in *Eurographics '91*, pp. 459–466. Amsterdam: North Holland, 1991.

[126] R. Siegel and J. R. Howell, *Thermal Radiation Heat Transfer*. Washington, D.C: Hemisphere Publishing Corp., 1981.

[127] P. Sloan, J. Hall, J. Hart, and J. Snyder. Clustered principal components for precomputed radiance transfer," in *SIGGRAPH*, 2003.

[128] P. Sloan, J. Kautz, and J. Snyder, "Precomputed radiance transfer for real-time rendering in dynamic, low-frequency lighting environments," in *SIGGRAPH 2002 Proc.*, pp. 527–536, 2002.

[129] I. Sobol, *Die Monte-Carlo Methode*. Berlin: Deutscher Verlag der Wissenschaften, 1991.

[130] M. Stamminger and G. Drettakis, "Perspective shadow maps," in *SIGGRAPH 2002*, pp. 557–562.

[131] L. Szécsi, "An effective kd-tree implementation," in Jeff Lander, editor, *Graphics Programming Methods*. Boston, MA: Charles River Media, 2003.

[132] L. Szécsi, L. Szirmay-Kalos, and M. Sbert, "Light animation with precomputed light paths on the GPU," in *GI 2006 Proc.*, pp. 187–194.

[133] G. Szijártó, "2.5 dimensional impostors for realistic trees and forests," in Kim Pallister, editor, *Game Programming Gems 5*, pp. 527–538. Boston, MA: Charles River Media, 2005.

[134] L. Szirmay-Kalos, *Monte-Carlo Methods in Global Illumination*. Institute of Computer Graphics, Vienna University of Technology, Vienna, 1999. http://www.iit.bme.hu/~szirmay/script.pdf.

[135] L. Szirmay-Kalos, "Stochastic iteration for non-diffuse global illumination. *Comput. Graph. Forum*, 18(3):233–244, 1999.

[136] L. Szirmay-Kalos, Gy. Antal, and B. Benedek, "Global illumination animation with random radiance representation," in *Rendering Symp.*, 2003.

[137] L. Szirmay-Kalos, B. Aszódi, I. Lazányi, and M. Premecz, "Approximate ray-tracing on the GPU with distance impostors," *Comput. Graph. Forum*, 24(3):695–704, 2005.

[138] L. Szirmay-Kalos, V. Havran, B. Benedek, and L. Szécsi, "On the efficiency of ray-shooting acceleration schemes," in *Proc. Spring Conf. on Computer Graphics (SCCG 2002)*, pp. 97–106, 2002.

[139] L. Szirmay-Kalos and I. Lazányi, "Indirect diffuse and glossy illumination on the GPU," in *SCCG 2006*, pp. 29–35.

[140] L. Szirmay-Kalos and G. Márton, "On convergence and complexity of radiosity algorithms," in *Winter School of Computer Graphics '95*, pp. 313–322, 1995.

[141] L. Szirmay-Kalos and W. Purgathofer, "Global ray-bundle tracing with hardware acceleration," in *Rendering Techniques '98*, pp. 247–258, 1998.

[142] L. Szirmay-Kalos, M. Sbert, and T. Umenhoffer," "Real-time multiple scattering in participating media with illumination networks," in *Eurographics Symp. on Rendering*, pp. 277–282, 2005.

[143] L. Szirmay-Kalos and T. Umenhoffer, "Displacement mapping on the GPU — State of the Art," *Comput. Graph. Forum*, 27(1), 2008.

[144] U. Thatcher, "Loose octrees," in *Game Programming Gems*. Boston, MA: Charles River Media, 2000.

[145] N. Thrane and L. Simonsen, "A comparison of acceleration structures for GPU assisted ray tracing," Master's thesis, University of Aarhus, Denmark, 2005.

[146] B. Tóth and L. Szirmay-Kalos, "Fast filtering and tone mapping using importance sampling," in *Winter School of Computer Graphics 2007, Short papers*, 2007.

[147] T. Umenhoffer, G. Patow, and L. Szirmay-Kalos, "Robust multiple specular reflections and refractions," in Hubert Nguyen, editor, *GPU Gems 3*, pp. 387–407. Reading, MA: Addison-Wesley, 2007.

[148] T. Umenhoffer and L. Szirmay-Kalos, "Real-time rendering of cloudy natural phenomena with hierarchical depth impostors," in *Eurographics Conf. Short Papers*, 2005.

[149] T. Umenhoffer, L. Szirmay-Kalos, and G. Szijártó, "Spherical billboards and their application to rendering explosions," in *Proc. Graphics Interface*, pp. 57–64, 2006.

[150] I. Wald, C. Benthin, and P. Slussalek, "Interactive global illumination in complex and highly occluded environments," in *14th Eurographics Symp. on Rendering*, pp. 74–81, 2003.

[151] I. Wald, T. Kollig, C. Benthin, A. Keller, and P. Slussalek, "Interactive global illumination using fast ray tracing," in *13th Eurographics Workshop on Rendering*, 2002.

[152] J. Wallace, K. Elmquist, and E. Haines, "A ray tracing algorithm for progressive radiosity," in *Computer Graphics (SIGGRAPH '89 Proc.)*, pp. 315–324, 1989.

[153] M. Wand and W. Strasser, "Real-time caustics," *Comput. Graph. Forum*, 22(3):611–620, 2003.

[154] R. Wang, D. Luebke, G. Humphreys, and R. Ng, "Efficient wavelet rotation for environment map rendering," in *Proc. 2006 Eurographics Symp. on Rendering*.

[155] A. Wilkie, "Photon tracing for complex environments," Ph.D. thesis, Institute of Computer Graphics, Vienna University of Technology, 2001.

[156] L. Williams, "Casting curved shadows on curved surfaces," in *Computer Graphics (SIGGRAPH '78 Proc.)*, pp. 270–274, 1978.

[157] M. Wimmer and J. Bittner, "Hardware occlusion queries made useful," in *GPU Gems 2*. Reading, MA: Addison-Wesley, 2005, pp. 91–108.

[158] M. Wimmer, D. Scherzer, and W. Purgathofer, "Light space perspective shadow maps," in *Eurographics Symp. on Rendering*, pp. 143–151, 2004.

[159] P. Wonka, M. Wimmer, K. Zhou, S. Maierhofer, G. Hesina, and A. Reshetov, "Guided visibility sampling," *ACM Trans. Graphics*, 25(3), 2006.

[160] A. Wood, B. McCane, and S.A. King, "Ray tracing arbitrary objects on the GPU," in *Proc. Image Vis. Comput. New Zealand*, pp. 327–332, 2004.

[161] S. Woop, G. Marmitt, and P. Slusallek, "B-kd trees for hardware accelerated ray tracing of dynamic scenes," *Proc. Graphics Hardware*, 2006.

[162] S. Woop, J. Schmittler, and P. Slusallek, " RPU: a programmable ray processing unit for realtime ray tracing," *ACM Trans. Graphics*, 24(3):434–444, 2005.

[163] C. Wyman, "An approximate image-space approach for interactive refraction," *ACM Trans. Graphics*, 24(3):1050–1053, July 2005.

[164] C. Wyman, "Interactive image-space refraction of nearby geometry," in *Proc. GRAPHITE*, pp. 205–211, December 2005.

[165] C. Wyman and C. Dachsbacher, "Improving image-space caustics via variable-sized splatting," Technical Report UICS-06-02, University of Utah, 2006.

[166] C. Wyman and S. Davis, "Interactive image-space techniques for approximating caustics," in *Proc. ACM Symp. on Interactive 3D Graphics and Games*, March 2006.

[167] J. Yu, J. Yang, and L. McMillan, "Real-time reflection mapping with parallax," in *Proc. Symp. on Interactive 3D Graphics and Games*, pp. 133–138, 2005.

[168] S. Zhukov, A. Iones, and G. Kronin, "An ambient light illumination model," in *Proc. Eurographics Rendering Workshop*, pp. 45–56, June 1998.

10489333R0

Made in the USA
Lexington, KY
28 July 2011